Professor Albert Einstein (Brown Brothers).

Understanding Einstein's Theories of Relativity

Man's New Perspective on the Cosmos

BY STAN GIBILISCO

DOVER PUBLICATIONS, INC., New York

This Dover edition, first published in 1991, is an unabridged, corrected republication of the work originally published by TAB BOOKS Inc., Blue Ridge Summit, Pa., 1983. The author has updated the list of "Suggested Additional Reading" for this edition.

Original edition edited by Roland Phelps

Manufactured in the United States of America
Dover Publications, Inc.
31 East 2nd Street
Mineola, N.Y. 11501

Library of Congress Cataloging-in-Publication Data

Gibilisco, Stan.
    Understanding Einstein's theories of relativity : man's new perspective on the cosmos / by Stan Gibilisco.
        p.   cm.
    Reprint. Originally published: Blue Ridge Summit, Pa. : Tab Books, c1983.
    Includes bibliographical references and index.
    ISBN 0-486-26659-1 (pbk.)
    1. Relativity (Physics)—Popular works.   2. Space and time—Popular works.
I. Title.
QC173.57.G5   1991
530.1′1—dc20   90-20570
                                                                CIP

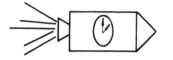

Contents

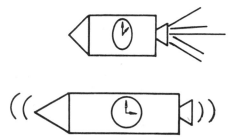

Introduction

THIS BOOK IS ABOUT THE STRUCTURE OF SPACE and time. It is about modern theories, as well as more ancient ideas, of the origin and constitution of our universe. It is about how time and space interact and behave. It is about where we came from and where we are going.

Astronomy has grown into a science of such magnitude and diversity that the average person today pretty much regards it as incomprehensible. The prevailing attitude is something like, "Leave it to the scientists to decide the fate of our universe." It is fascinating just to look through a small telescope on a clear night and see the craters of the moon, the rings of Saturn, a star cluster, or another galaxy. But modern astronomy is much more than that. The very fiber of the universe is being explored; theories are being spun that would make the most conservative ideas of the ancients seem more comprehensible. It is wilder than any science fiction!

Many cosmologists—those who study cosmology, the structure of the universe—now believe that space, while having no boundaries, has only a finite size. Others have seriously suggested that matter is being formed continuously from nothing, all over the universe. It has been theorized that there are places in space where gravity is so intense that even light cannot escape, and anything that gets too close will be irretrievably sucked in and squeezed to unreal proportions. It is quite possible that our universe is connected with others, via passageways where time is stopped and matter is reduced to geometric points.

Where does all this lead? Perhaps we ought just to go back to the idea that all the stars are holes in a big black sphere surrounding the earth. Maybe we should not ask what might lie beyond this sphere. Perhaps we should save our mental energy for more important things. But, after a while, such an attitude becomes boring.

Cosmology is within the grasp of everyone. We should, as a scientifically oriented society with the good fortune of an abundance of leisure time, keep reasonably abreast of the developments in astronomy and cosmology. Even without the mathematical training necessary to follow every detail, anyone can get sufficiently acquainted with cosmology to ask intelligent, probing questions. It is a science of more questions than answers.

Here, we will begin by explaining the development of the special theory of relativity and progress through the general theory of relativity into modern cosmology. There are some mathematical formulas and derivations, but they can be fully understood by anyone with a high-school level of training. There are many illustrations. Quite often, you will be asked to follow along and imagine things—sometimes quite improbable things! These little "mind journeys" are helpful to the understanding of relativity and cosmology.

The special theory of relativity is concerned with the effects of the constancy of the speed of light (c) on the behavior of moving objects in space. The theory of general relativity carries this further and involves the structure of the universe as a whole. Beyond this, we will look at theories such as the big-bang and steady-state models of our universe and the fascinating theoretical concept of black holes. All you need is a vivid imagination and the willingness to use it.

Chapter 1

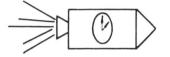

The Speed of Light
Is the Speed of Time

I N THE LATE PART OF THE NINETEENTH CENTURY, physics was an unpopular field. It was regarded as a science that had reached total maturity; everything there was to know about it had been discovered. It wasn't a challenge anymore. With no hope for further adventure, physics was dead—an intellectual graveyard.

Then, in the early years of this century, some brilliant scientists revolutionized physics and astronomy by disproving the accepted Newtonian principles. At this time, the new science of cosmology was born. Suddenly, it seemed that hardly anything was known about the subject. The most famous of these pioneers was Albert Einstein, and his model of the universe became known as the theory of relativity.

THE FUNDAMENTAL AXIOM OF RELATIVITY

The premise on which Einstein's entire theory is based is that the speed of light, and all electromagnetic radiation, is always the same, no mat-

ter from what point of view it is measured. In free space, this speed is about 186,000 miles, or 300,000 kilometers, per second. At such a speed, a ray of light travels from the sun to the earth in about eight minutes. The light of the moon takes only a little more than one second to get to our planet.

The significant consideration is that this speed is always the same. It would appear identical whether we measured it on the earth, the moon, the sun, on board a space vessel, or in an intense gravitational field.

What caused Einstein to postulate this? Perhaps it was his stubborn belief in an orderly universe; perhaps he thought such a principle was mathematically elegant because of its simplicity. Whatever his reason, he set out to discover, mathematically, in the privacy of his own mind, things that have affected the course of human history to an extent previously unparalleled.

As an example of this principle, imagine that we are on board a hovering spaceship somewhere

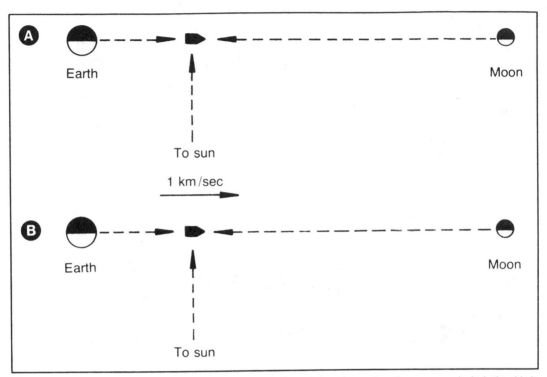

Fig. 1-1. Measuring the speed of light from different moving points of view always gives the same result. At A, the ship is hovering above the earth; at B, it is moving away from the earth toward the moon at 1 kilometer per second.

between the earth and the moon, as shown in Fig. 1-1A. Suppose an earth-based space station shoots a laser beam out toward us. Imagine further that we are equipped with very complex and sophisticated apparatus that enables us to measure the speed of this beam with extreme accuracy. We measure the speed of the arriving laser beam and find it to be 299,792.50 kilometers per second. We measure the speed of the light arriving from the sun and find that it is also 299,792.50 kilometers per second. We measure the speed of the moonlight, and obtain, once again, the same value.

Now, suppose we ignite our engines and accelerate our vessel to a velocity of 1 kilometer per second directly away from the earth, as shown in Fig. 1-1B. Focusing our speed-measuring equipment on the laser beam from the earth-based observatory, we measure the speed of the arriving light. You might expect it to be 299,791.50

kilometers per second, exactly 1 kilometer per second less than previously. Since we are retreating from the source, such a conclusion appears logical, but no—it is not any slower—it is exactly the same. Turning toward the sun and then the moon, we find that the light arriving from them is still coming in at exactly 299,792.50 kilometers per second. Why should this be?

This effect has been experimentally verified now that we have equipment capable of measuring the speed of light very accurately. It is no longer just a theory. In the early twentieth century, there was no positive certainty about the truth of Einstein's fundamental axiom. Now, we know it is true.

REFERENCE FRAMES

An observer's perception of the universe depends on his point of view, or reference frame. A reference frame may be stationary, moving at a given speed in a given direction, or accelerating at a given rate in a given direction. However, in order to make any sense, motion (or lack of motion) must be measured with respect to another object. We cannot simply say that something is moving at x meters per second; we have to specify a reference frame to define motion.

Consider the simple example of a universe with only one object. That is, suppose there is a continuum that is absolutely empty except for one piece of matter. Imagine that you are that object, floating in the total blackness of an otherwise unoccupied universe. How can you say whether you are moving or rotating or stationary? All you can perhaps surmise is that, since there appear to be no forces pulling in any direction on your body, you are not accelerating. Otherwise, you have nothing with which you can compare your reference frame. Motion is without meaning.

Now, imagine that you are equipped with a gun of some kind, still floating in the otherwise empty universe; suppose this gun has large, heavy bullets that generate considerable recoil when the gun is fired. Suppose this gun propels its bullets at a speed of 1,000 meters per second.

After firing the gun one time, you can reasonably conclude that its bullet is traveling through the void away from you at 1,000 meters per second. Has its recoil effect set you in motion? This you cannot say, since the only other object you have to compare your motion with is the bullet. No absolute reference frame exists. It is just as logical to base the motion in this universe on the bullet as it is to base it on yourself; you are moving away from it, and it is moving away from you (Fig. 1-2A).

But, now suppose that you fire the gun at intervals, in the exact same direction, each time causing some recoil. For convenience, imagine that the gun is fired at exact one-second intervals. Now, you can suppose that each bullet gives you a little more push, causing you to accelerate in bursts, in the opposite direction from that of the stream of bullets. This is shown by Fig. 1-2B. What reference frame can we use to determine whether or not your body is, in fact, accelerating? A good choice would be the first bullet. If you measure your speed with respect to the first bullet, you will find that your speed increases slightly each time you fire the gun. Your velocity will always be 1,000 meters per second away from the most recently fired bullet; compared to earlier and earlier bullets, your speed will be greater and greater.

Under these circumstances, the only objects available for ascertaining your motion in the universe are the bullets, unless you assume (postulate) that you are the absolute standard of reference in your universe. This might be a good assumption. It is certainly tempting to postulate this in an effort to make matters simple. Unfortunately, doing this leads to an inability to explain certain effects.

Remember that the gun always fires each bullet with a speed of exactly 1,000 meters per second. If you consider yourself as always stationary, then all bullets fired from the gun should always move through the void away from you at exactly that

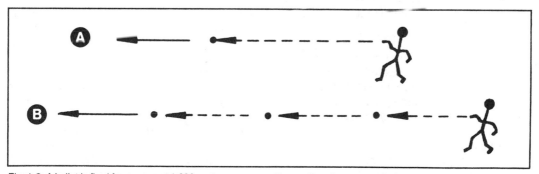

Fig. 1-2. A bullet is fired from a gun at 1,000 meters per second in an otherwise unoccupied universe. At A, a single bullet is fired; at B, several bullets are fired in succession.

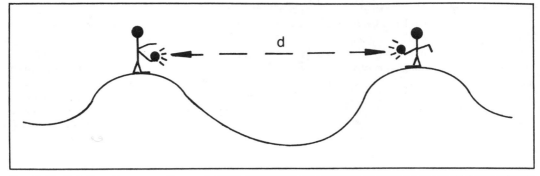

Fig. 1-3. The first attempt to measure the speed of light. As far as the experimenters could tell, the speed of light was infinitely fast.

speed. This should be true no matter how many times you fire the gun. However, more distant bullets appear to move faster; the earlier a bullet was fired (that is, the more bullets you shoot after a given bullet), the greater its velocity will be. It would be very difficult to explain why the firing of the gun would cause an entire stream of distant bullets to speed up! And all by the same amount!

The examples just given are somewhat unrealistic, since of course, we do not exist in an otherwise empty continuum. Our universe has trillions of objects. They are all moving with respect to each other in an infinite combination of velocities. Finding any one object to call "at rest" is simultaneously trivial and impossible. Every object is, of course, stationary with respect to its own reference frame, and in motion relative to most other objects. Relativity theory asserts that every point of view is as good as every other; there is no absolute standard of motion in the universe.

LIGHT SPEED

The idea that the universe has no absolute standard for motion is consistent with the fundamental axiom of relativity. No frame of reference has any advantage with respect to the speed of light or absolute motion. All points of view in the universe are equivalent.

All electromagnetic radiation travels with the same velocity as light. Relativity theory goes further, to state that speeds greater than light are not possible, and that all forces and effects, as well

as material objects, are limited to this speed. There is no known way to transfer information via any medium that goes faster.

How can we measure the speed of a light beam? The earliest attempts were conducted long ago between experimenters with lanterns. Their method is illustrated by Fig. 1-3. By prior agreement, they arranged to have one man uncover his lantern, and the other man was to uncover his lantern as soon as he saw the light from the first. The first man would then ascertain the period between the time he uncovered his lantern and the moment he saw the light from the second. This experiment was conducted over distances as large as practicable, such as between two hilltops. Needless to say, the only significant delay in transit was the result of the imperfection of the reflexes of the second man.

Later attempts to measure the speed of light proved more fruitful. It is now possible to measure the speed of light with fair accuracy using equipment found in a high-school physics or electronics laboratory. Figure 1-4 shows the basic arrangement.

The light source, consisting of a neon lamp capable of being modulated at a rapid rate, is driven by a series of pulses with waveforms similar to those in Fig. 1-5A. The light from the bulb is collimated by means of a lens, and the resulting beam is aimed at a mirror several meters away. The mirror reflects the beam back to the photoelectric cell, which is oriented so that it receives some light directly from the bulb as well as the light reflected from the mirror. The photoelectric cell is connected to an oscilloscope as shown in Fig. 1-4.

The oscilloscope displays two series of pulses, as shown in Fig. 1-5B. One series is the result of the light arriving directly from the bulb, a few inches away; the other is the result of the light reflected off the mirror. The latter beam has traveled a considerable distance. By adjusting the orientation of the photoelectric cell, lens and bulb so that both pulses are of equal amplitude on the scope, the time delay can be measured. Then, knowing the difference between the distances traveled by the light beams causing the two sets of pulses, the speed of light can be calculated. Such a method is satisfactory for measurement to within a few percent.

Once the technology had been obtained to discover that light did not, in fact, travel instantaneously from place to place (as had been previously thought), scientists naturally wondered whether the speed of light would be affected by motion. They knew that the speed of sound was affected by motion, and they theorized that light must also be influenced. They postulated that light, in order to travel, must be carried by a medium of some kind, just like sound. This idea became known as the *ether theory*—the medium for light propagation must be present everywhere, even in a total vacuum.

Perhaps this ether was the key to the absolute motion standard for the universe. If light traveled with constant speed in the ether, then we ought to be able to determine, by measuring the speed of light in various directions, the speed of the earth with respect to this ether. Such was the reasoning at the time.

ETHER THEORY

We know that sound travels by conduction, with a fixed speed, through the atmosphere. If a source of sound waves is stationary relative to the surround-

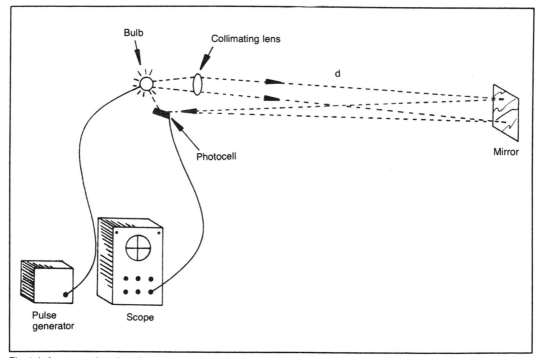

Fig. 1-4. A more modern, but elementary, method of measuring the speed of light. The path distance is d. This apparatus can be set up in a high-school physics lab.

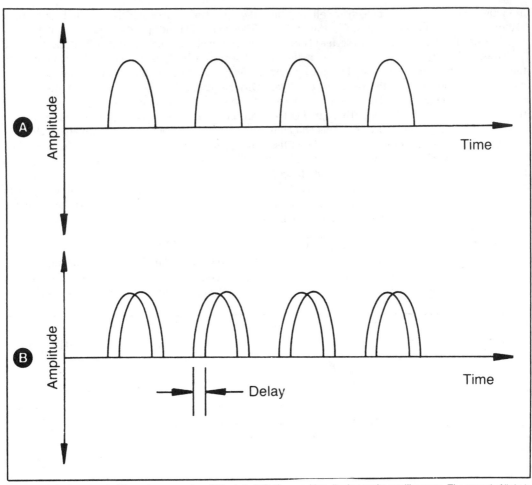

Fig. 1-5. At A, the output of the pulse generator of Fig. 1-4. At B, the resulting display on the oscilloscope. The speed of light is the distance d divided by the delay time.

ing air, then the speed of sound is the same in all directions away from the source (Fig. 1-6A). If the source is in motion relative to the surrounding air, then the sound travels move rapidly away from the source in the direction opposite its motion, and more slowly in the direction of its motion (Fig. 1-6B). The speed of sound is constant with respect to the air, regardless of the motion of its source. Would the same be true of light? Experiments were carried out to see.

Figure 1-7 shows one setup to measure the possible effect of the motion of the earth relative to the other. Light from a distant star was measured at two times: first, when the earth was moving toward the star in its orbit around the sun, and six months later, when the earth was moving away from the star. It was hoped that this and other experiments would provide a clue as to the absolute motion of the earth with respect to the universe as a whole; for, of course, the ether should be at rest with respect to the universe, constituting an absolute standard of motion.

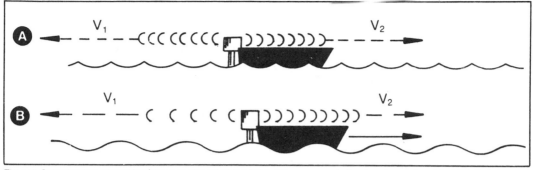

Fig. 1-6. Sound waves emanating from a stationary object (A), in this case a motorboat, and a moving object (B). At A, $v_1 = v_2$, but at B, $v_1 > v_2$, where velocities are measured relative to the sound source.

Every experiment to detect the motion of the earth with respect to the conductive ether turned out negative. Scientists developed elaborate explanations to preserve the idea that there might after all be some kind of conductive medium in the universe, but it simply was not consistent with observed fact. The main objective of the ether theory was to determine an absolute standard of motion for the universe; by defying all experimental efforts to find such a standard, the ether theory was eventually scrapped. Some other explanation for the constancy of the speed of light propagation had to be found.

INDIRECT OBSERVATIONS

Under certain circumstances, the speed of light may not appear constant. However, such instances always involve measurement from a remote point, rather than in the vicinity of the light beam itself. One such set of parameters is shown in Fig. 1-8.

Imagine that there are three space stations arranged as illustrated, at the vertices of a right equilateral triangle. A laser beam is fired from station A toward station B. Since the distance separating A from B is 300,000 kilometers, light will complete the trip in one second. If we place a mirror at station B and orient it so the light beam is reflected back to A, an observer at A will measure a two-second delay and conclude that light travels at 300,000 kilometers per second.

But, suppose we are observing this from station C. Imagine that there is a lot of interstellar dust

between A and B, so that we, at C, can watch the ray of light as it moves across the space between stations A and B. In this case, the beam will seem to complete the transit in 1.414 seconds, not one second. The beam will appear to move more and more slowly as it gets further and further from station A.

The reason for this apparent paradox is that the image of the ray of light, projected onto the interstellar dust, must travel greater and greater distances to our observation point at station C. This image itself can, after all, travel only at the speed of light. Upon first leaving station A, the image travels 300,000 kilometers, or one light second. By the time it reaches station B, the ray of light is $300,000 \times \sqrt{2}$ kilometers, or 1.414 light seconds, from us; the image consequently requires 1.414 seconds to arrive at our observation point. We see the beam leaving A as it is really just arriving at station B, and we see it reach B $\sqrt{2}$, or 1.414, seconds after that.

Suppose that the light beam is now diverted just a little, so it misses station B and goes on into space. As the light beam travels past B, it will appear to move more and more slowly. If we move station B out to greater and greater distances, the discrepancy between the actual and observed transit time will get greater and greater. The discrepancy will approach a factor of 2 as the distance is increased without limit, so that if the distance is very great, the beam will appear to take nearly twice as long to travel the path, as compared to its actual transit time.

Figure 1-9 shows an extreme example. Here,

stations A and B are 100 light seconds, or 30,000,000 kilometers, apart. We see the light ray leave station A when it has traveled one light second toward B. The beam actually gets to station B 99 seconds after that. But the image will not arrive at our observation station until a little more than 100 seconds later still. Thus the beam will seem to require a bit over 199 seconds to make the trip over a distance that is actually 100 light seconds!

If we reverse this situation—that is, shoot the beam from station B to station A—the beam will appear to travel almost infinitely fast. This is illustrated by Fig. 1-10. In this situation, the apparent transit time will be only 0.995 seconds. We see the

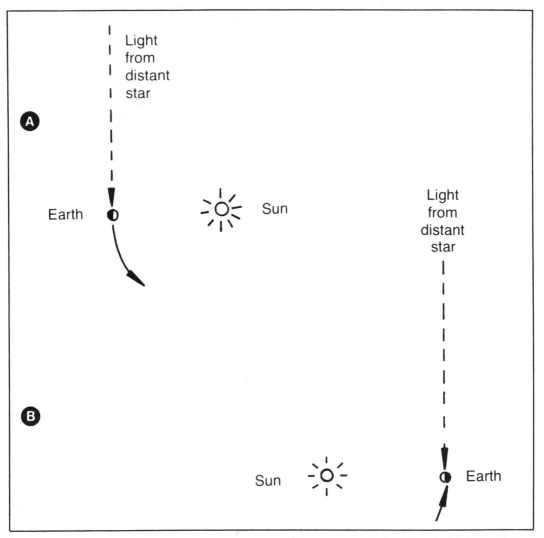

Fig. 1-7. Experiment for determining the effect, if any, of the ether on the apparent speed of light. At A, the light from a star is put through speed-measuring equipment as the earth is moving away from the star. At B, the test is repeated six months later, when the earth is moving toward the star.

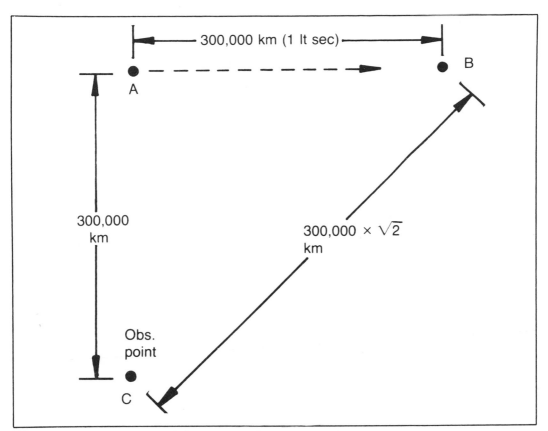

Fig. 1-8. The projected image of a light beam may not always appear equal to c.

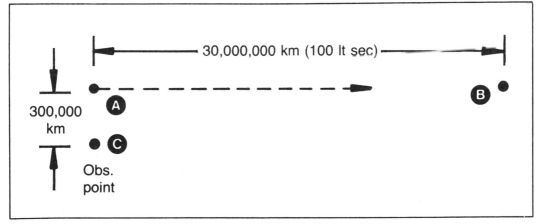

Fig. 1-9. A more extreme example of the apparent speed change of a projected light beam.

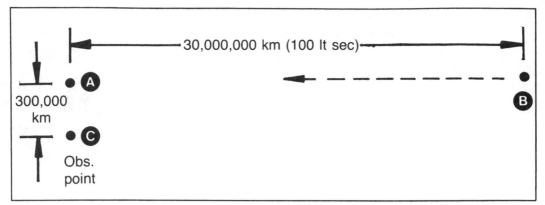

Fig. 1-10. Apparent speed increase of a projected light beam traveling on an approaching path.

beam leave station B after it has actually arrived at A! The image of the ray travels almost the same path through space as the ray itself.

These examples may at first seem to violate, or invalidate, the fundamental axiom of relativity. To keep the theory consistent, we must amend the axiom to state that the speed of light is always constant, no matter what the reference frame from which it is measured, as long as we measure its speed *directly* (such as with the method of Fig. 1-4). If we measure the speed of an *image*, we will not necessarily get an accurate indication, since the image itself can travel only at the speed of light, and thus needs a certain amount of time to reach us.

THEORETICALLY FASTER THAN LIGHT

A consequence of the fundamental axiom of relativity is, as we shall see in later chapters, the fact that we cannot get any material object to travel at the speed of light. No matter how much energy is expended to accelerate a piece of matter, the resulting speed will always be slower than the speed of light. No matter how powerful our spacecraft engines ever get, we will never be able to travel at the speed of light. The speed of light is the fastest possible velocity in our universe.

A statement such as this will surely invite attempts to disprove it. One of the most popular arguments, or models, used to try to invalidate this is shown in Fig. 1-11. Suppose we have two rigid rods, immensely long, with absolutely no elastic-

ity. This may allow us to exceed the speed of light in a certain theoretical sense. If we hold the two rods so they cross a few feet in front of us (A), and then slowly move them toward a parallel orientation, the intersection point of the two rods will move away from us at an ever-increasing rate (B). If the rods are long enough, the point of intersection will move away in an asymptotic manner, whereupon the intersection will suddenly disappear as the rods become parallel. At some time just prior to the rods becoming parallel, the velocity of this intersection point will surpass the speed of light. This is necessary in order that the point can travel an infinite distance in a finite amount of time.

Of course, such a model is impractical for several reasons. Such rods cannot be constructed; there is no such thing as a total lack of elasticity, and the length necessary to achieve the extreme speed would be ridiculous. But even if we could obtain such rods, how do we know that the impulse of their changing orientation can travel faster than light? Might the rods actually bend because such speed simply cannot happen? The fundamental axiom of relativity would seem to say that this would in fact be the case. Even the effect of changing orientation should obey this law. Unfortunately there is no experimental way to determine this.

We can substitute beams of light for the rods, and perhaps conduct the experiment in a realistic way. Light always goes in a straight line, and it is a trivial task to change the direction of a collimated light source. Figure 1-12 illustrates how we might attempt to generate speeds faster than light, if only for a theoretical point of intersection. While this point is not a material thing, it does exist in the sense that, at the point, light is shining from both sources, and this is true of no other point in space.

Suppose, then, that these beams cross at a certain angle and we suddenly orient the beams so that they are parallel. In theory, the intersection point should move away with extreme speed. In fact, the point will move away in a manner shown by Fig. 1-13. After, say, one second, the beams will be parallel out to a distance of 300,000 kilometers, but will remain in their original positions at greater distances. The light requires time to travel, and until the light emitted by the lasers at their new positions reaches a certain place in space, the rays will appear as they were before the change. After

two seconds, the beams will be parallel out to 600,000 kilometers; after ten seconds, out to 3,000,000 kilometers, and so on. The crossing point will be limited in its speed because of this distortion. This is a case where the light actually does not go in a straight line because of the change in direction of the sources.

Still another example, once again involving objects that cannot really exist but are useful to imagine for illustrative purposes, is the rope-and-pulley model. Imagine that we string a pulley system between the earth and the moon, as shown in the diagram of Fig. 1-14. Suppose that this rope is completely inelastic, so it will not stretch no matter how long it is. If we pull on one end of the rope, will the other end move back immediately, or will there be a time delay?

The moon is approximately 400,000 kilome-

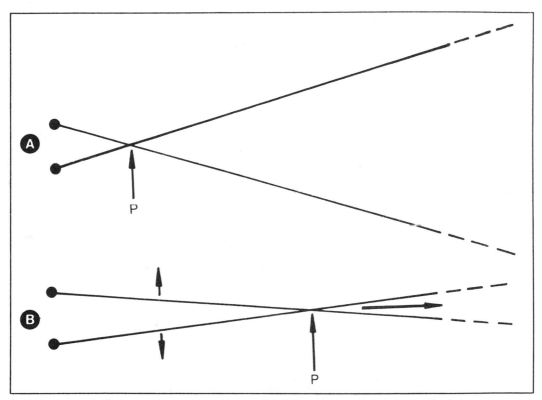

Fig. 1-11. A theoretical attempt to generate speeds faster than light using two extremely long, stiff rods.

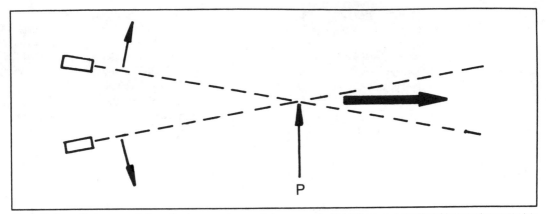

Fig. 1-12. A practical means of attempting to generate speeds in excess of c. Point P may eventually achieve such speeds, it is hoped.

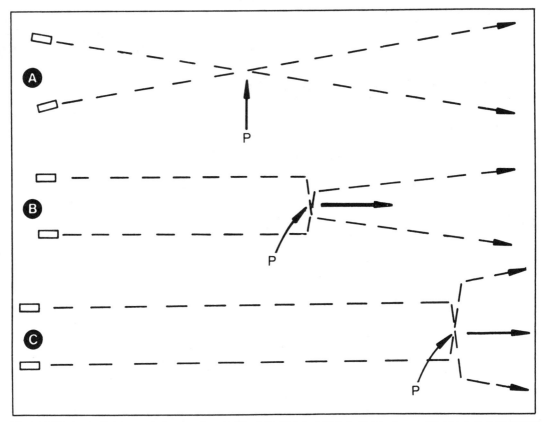

Fig. 1-13. What actually happens in the experiment of Fig. 1-12. At A, the initial situation is shown; at B, after a short interval of time; at C, after a longer time.

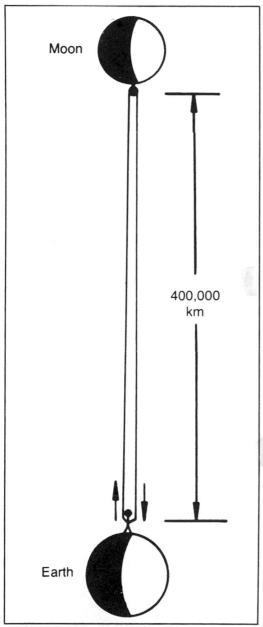

Moon

400,000
km

Earth

Fig. 1-14. The rope-and-pulley experiment. A non-elastic rope is strung between the earth and the moon, a total round-trip distance of 800,000 kilometers. Then one end of the rope is pulled, and the other end retracts. The speed of the impulse is measured.

ters from earth. The impulse must therefore travel 800,000 kilometers. This is roughly 2.7 light seconds. Relativity theory asserts that the retraction will be delayed by this amount of time because the impulse—the fact that the rope was pulled—can itself only move at the speed of light.

How can the speed of light be so absolute? Relativity theory simply postulates that it is. A second axiom mandates that nothing can go faster than light: not material objects, not gravity, not electric or magnetic forces, nor even facts or ideas! In this sense, the speed of light is the speed of time.

We now have two postulates, asserted as absolute truths without proof. We have experimentally verified the fundamental axiom of relativity since Einstein first proposed it; there may still come the time when a contradiction is derived from the axiom, but as yet this has not happened. In order to develop a scientific theory, we must have such absolutes, and then make deductions based on them. Of course, we should not indiscriminately postulate something just because it sounds good; doing this would certainly result in logical contradictions somewhere. If we run across a contradiction in a theory, it means the theory is inconsistent, and we had better change the set of axioms. As yet, this has not occurred in the theory of relativity.

SEEING BACK IN TIME

What if we could violate this second axiom somehow, and go at speeds faster than light? What might we be able to do? Here is one example.

Imagine that we have extremely powerful telescopes, and very sensitive radio receiving equipment, on board a spacecraft that can travel twice the speed of light. Suppose we get on board that ship and travel a distance of two light years directly away from the earth at the maximum speed of 2c. By doing this, we will be able to "catch up" to light that left the earth before we did. We will not only catch up to the light that left the earth, but we will also catch up to all the electromagnetic radiation that was emitted, such as from television and radio broadcasting stations. Moving at 2c for a distance of two light years, we could catch up to the light that left the earth a year before we left.

Focusing our super telescope on the home planet, we would be able to see events in the past; we might even glimpse ourselves getting out of our house and heading for the car to go to the drug store to get some aspirin for that terrible headache. With our radio receiving equipment, we might listen to last year's football games all over again. With television, we could have a complete replay. But, these would be no tape recordings; we would be seeing the actual events and listening to the actual signals!

There is a certain intuitive uneasiness in imagining things like this. We could actually broadcast a message ourselves, and then go listen to it and perhaps watch our actions. It would be like existing in two places at the same time.

Continuing further out into space at the speed of 2c, we could listen to the very first wireless broadcasts; after that we would no longer hear anything on the airwaves because the first experimental broadcasts had not yet reached us. In theory, with a powerful enough telescope, we could watch our grandparents playing in their yards; we could watch the Revolutionary War; we might even find out how man evolved.

But relativity theory denies all this. It can only remain in our imaginations. We'll never do it. We cannot travel any faster than light, at least in our universe. Perhaps there are other universes, and mathematically there may be ways to get around this second axiom of relativity in an indirect way. We will have more to say about this in later chapters.

WHAT IS LIGHT?

We have seen that light, and all electromagnetic energy as well as electric and magnetic fields, travel at a constant speed through the universe, and that this speed in vacuum is the same as measured from all non-accelerating reference points. But what is electromagnetic energy? How do we explain its propagation? Scientists have developed two models to explain the behavior of electromagnetic propagation, and these theories are commonly called the "particle theory" and the "wave theory." These are self-explanatory terms. Light displays certain properties that make it seem like it is composed of particles. In certain other ways, it behaves like a wave. These two models of electromagnetic energy, both with merit, make it difficult for us to say exactly what light is, just as atomic theory makes it quite a task to explain exactly what matter is. Here, we will briefly describe what we can observe about electromagnetic energy, consistent with these two models.

PARTICLE THEORY

Scientists first developed the idea that light might consist of discrete corpuscles in the time of Isaac Newton. Today, we notice that light behaves as a stream of particles, known as quanta or photons, which represent the smallest possible packet of electromagnetic energy. As such, they are indivisible. You can't have part of a photon.

To illustrate quantized light, imagine the experiment of Fig. 1-15. Suppose you shine a flashlight onto a wall for exactly one second. The light striking the wall may consist of, say, one joule of energy if the bulb emits one watt of light. Now, imagine that you shine the light for only 0.5 seconds; then only 0.5 joules of energy is intercepted by the wall. Suppose you keep cutting the length of time in half: You switch the light on for ¼ second, ⅛ second, and so on, halving the amount of energy again and again. Theoretically, you can keep doing this forever, neglecting of course the limits imposed by your reflexes and the nature of the physical apparatus. You can always divide a positive, nonzero number by two, and get another positive, nonzero number.

According to the *corpuscular theory* of light, however, you cannot continue this process forever. After a certain number of repetitions, you will reach a point where just a few photons, perhaps eight of them, strike the wall. On the next flash, half as long, four photons will hit the wall, and then two, and finally only one. What after that? You cannot have half a photon!

On the next flash of the light, all we can say is that we might get a photon and we might not, and that the probability is 50 percent that we will. On the next flash, the probability is 25 percent that a photon will strike the wall; after that, 12.5 percent, and so on. Of course, there might be extraordinary occurrences where we get two photons that happen to be emitted almost simultaneously, but the point is that we will never, ever get part of a photon.

Of course, no one has ever done this particular experiment and gotten these results. But scientists have seen that certain atomic reactions produce energy by-products in discrete amounts. They have

also actually observed that a beam of light puts physical pressure on any object it shines upon.

The amount of energy carried by a single photon can vary. A photon of red light carries less energy than a photon of green light, which in turn carries less energy than a photon of violet light. Knowing the amount of energy carried by a photon, we can determine the kind of energy. Figure 1-16 illustrates the relation between the energy in a

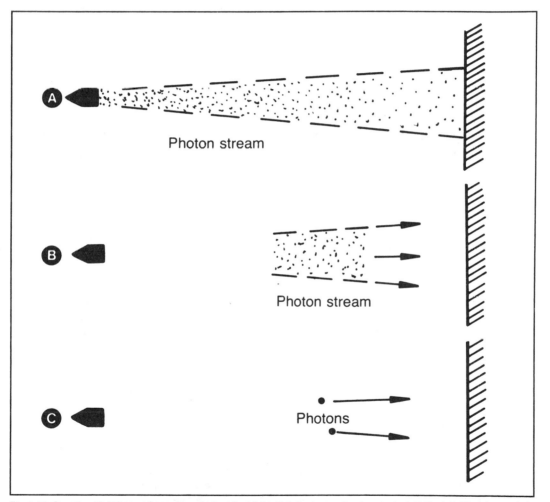

Fig. 1-15. Quantized light beams from a hypothetical flashlight. At A, the beam at a given instant. At B, a brief flash is caught in mid-flight. At C, an extremely short flash produces only two photons.

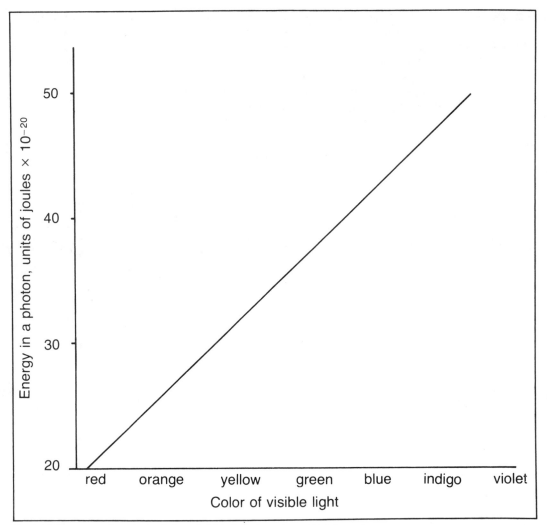

Fig. 1-16. The energy in a single photon of visible light depends on the color.

single photon and the color of the light beam resulting from a barrage of these photons.

A single photon carries only a minute amount of energy. Mathematically, this energy is on the order of quintillionths of a joule for visible light, septillionths or octillionths of a joule for radio signals, and perhaps a few trillionths of a joule for X-rays. But for a given type of electromagnetic energy, all the photons carry the same amount of energy, and it is not possible to have any quantity of energy not a multiple of this amount.

The photons comprising any electromagnetic energy travel, of course, at the speed of light. The particle theory of light is very convenient for answering the question, "Why does light travel at finite speed and not infinitely fast?" But when we ask, "Why the particular speed of just under 300,000 kilometers per second in a vacuum?" the particle theory is not of much use. Intuitively, the particle theory might tempt us to believe that photons carrying greater amounts of energy (such as blue light) should travel faster than photons with less energy (such as red light); but they all move in a vacuum at the same speed. The photons with

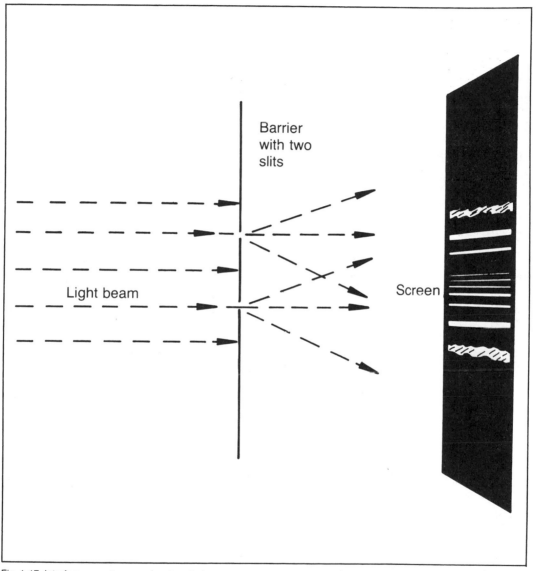

Fig. 1-17. Interference patterns are produced when monochromatic light is passed through two narrow slits. This shows that light has wave-like properties.

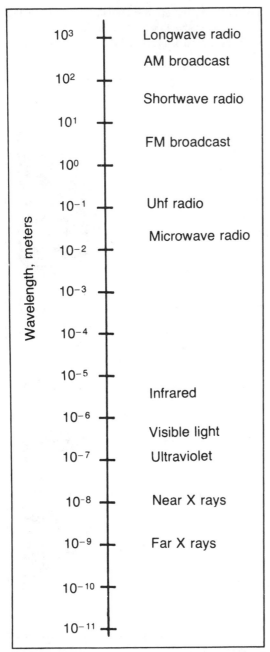

Wavelength, meters	
10^3	Longwave radio
	AM broadcast
10^2	
	Shortwave radio
10^1	
	FM broadcast
10^0	
10^{-1}	Uhf radio
	Microwave radio
10^{-2}	
10^{-3}	
10^{-4}	
10^{-5}	
	Infrared
10^{-6}	
	Visible light
10^{-7}	Ultraviolet
10^{-8}	Near X rays
10^{-9}	Far X rays
10^{-10}	
10^{-11}	

Fig. 1-18. The electromagnetic spectrum from a wavelength of 1,000 meters to 10 trillionths of a meter. (This is a logarithmic scale.)

more energy have more mass. In that sense, they are heavier, and cause greater pressure when they strike a barrier.

Mathematically, the relation between the mass of a photon and the energy it carries is given by the formula:

$$E = mc^2$$

where E is the energy and m is the mass, in joules and grams respectively. The constant in this equation is c, the speed of light.

WAVE THEORY

Electromagnetic energy shows wave-like behavior under the right conditions. This can be easily demonstrated by passing a beam of light through two narrow slits spaced very close to each other. The resulting interference pattern gives strong evidence to the idea that light is a wave (Fig. 1-17). Light waves, like ripples on a pond, tend to bend around sharp corners and diffract through narrow slits. In fact, "ripple tanks" are used quite often to demonstrate wave behavior.

The wave theory of light gives rise to the thought that there must be some kind of conductive medium to carry electromagnetic energy. This theory is not, however, consistent with other observed facts. How do waves travel through a total void? Apparently the simple existence of electric and magnetic forces is sufficient; the idea that there must be some kind of medium to carry these forces is a notion that appeals to our intuition but need not necessarily be a fact.

All waves result from oscillations. Electromagnetic waves are caused by the vibration of electrically charged particles. Wave motion displays two properties, frequency (represented by f) and wavelength (represented by the Greek letter lambda λ). These two properties of a wave are related to each other by the constant factor c:

$$c = \lambda f,$$

where λ is specified in meters, f in oscillations per second, and c in meters per second.

Radio signals have a wavelength that can be

measured in meters; visible light waves must be measured in millionths of meters, or microns; they are often also measured in units of 10^{-10} meters, known as Angstrom units. Electromagnetic radiation can occur at any wavelength. Figure 1-18 shows the electromagnetic spectrum, and the wavelengths of various kinds of electromagnetic energy.

Electromagnetic waves travel more slowly in substances such as water, as compared to their speed in a perfect vacuum. Hence the speed of light, c, is less in such substances. Moreover, the speed of light in these substances varies with the wavelength! This puzzling effect mandates a further refinement of the fundamental axiom: We must restrict our measurement of the speed of light to the same medium, and the same wavelength if the medium is not a vacuum, in order for c to be constant under all conditions. In all discussion henceforth, we will be considering the medium to be a perfect void, eliminating these complications.

The wave theory and the particle theory of light are quite different models for the same phenomenon, and this might lead you to wonder whether scientists really know what light is. In fact, about all we can say to answer that question is that light is energy. It displays some properties of both a particle and a wave. Sometimes the particle model provides a better explanation of observed data, and sometimes the wave model seems more descriptive.

MATTER IS ENERGY

Einstein's equation, $E = mc^2$, gives us one of the most important results of relativity theory. Matter and energy are actually different manifestations of the same thing. We may very well say, then, that light is matter! Matter can be converted into energy; the most common example of this is the atomic bomb. As yet, we haven't found a way to convert energy back into matter, but perhaps some day this will also be possible. To find the amount of energy that can theoretically be derived from a given mass, we multiply the mass by the constant factor c^2. To determine the mass equivalent of a given amount of energy, we divide by c^2. The speed of light, c, defines the relation of matter and energy.

The total amount of matter/energy in the universe may or may not be constant. If the amount is changing, it may enter or leave our universe in either form. Some older models of the universe consider it to be finite, but surrounded by a void that extends on without end; the universe therefore continuously releases energy into this void, its ultimate fate being a cold, dark end. Some more modern models provide for the continuous addition of matter, being formed everywhere from nothing. Some cosmologists believe that the total amount of matter/energy in our universe is finite, but the universe has no boundaries and yet is folded over onto itself so that the unpleasant slow death by continuous energy loss is avoided. We will examine these pictures of our universe in Chapter 9.

TIME IS DISTANCE

The constant c also provides us with a means of defining time in terms of spatial separation, and vice-versa. The relation is simply:

$$d = tc$$

where d represents distance and t represents time.

According to this definition, we can define a span of 300,000 kilometers as being a manifestation of time equivalent to one second, we can also define one second as a spatial separation of 300,000 kilometers. This concept allows us to visualize time as a fourth dimension, through which our universe is hurtling with velocity c towards the future! This model can be used to good advantage for illustrating certain space-time effects.

Figure 1-19 shows how a two-dimensional continuum would move through a third time dimension with speed c. Any two events that occur 300,000 kilometers apart—regardless of the orientation of the line connecting them—are separated by a time difference of one second. The converse is also true; any two events one second apart in time are also precisely 300,000 kilometers apart in distance.

The model of Fig. 1-19 will be used in greater measure when we discuss polydimensional geometry. Imagining our universe to be moving

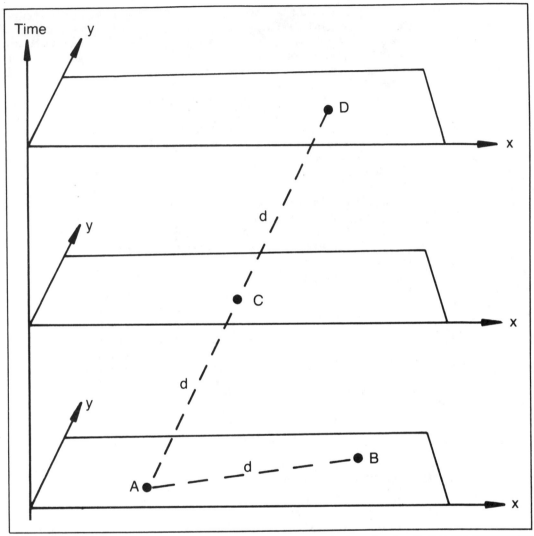

Fig. 1-19. Space-time model for a plane continuum. The plane moves upwards at speed c, along the time axis. Any two points separated by distance d are one second, or 300,000 kilometers, apart, regardless of the actual positions of the points.

through a fourth time dimension, at velocity c, is only a mathematical illustration, showing that the speed of time and the speed of light can be considered equivalent. Although this has the effect of reducing the whole idea to a tautology—nothing more than a mathematical construction—the model clearly demonstrates, as we shall later see, very real effects.

Chapter 2

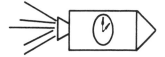

What Does "Simultaneous" Mean?

W HEN WE SAY TWO EVENTS OCCUR SIMULTANE-
ously, we supposedly mean that they happen at the same time. This concept could hardly be
simpler in the everyday world; and yet, on the
grander scale of interplanetary and interstellar
space, this simple idea becomes complicated.
Under certain conditions it becomes totally meaningless.

TIME STATIONS

You have no doubt heard the expression, "Synchronize your watches!" You can set them to read
exactly alike; with the new digital watches, you can
adjust two of them so they agree within the limit of
your reflexes—a few hundredths of a second. You
might set them by the National Bureau of Standards
time station, WWV. If your reflexes are quick
enough, you can set two watches so closely that
they both change from "59" to "00" seconds at the
same instant the tone occurs to indicate the start of
a new minute. Or, at least, as closely as your eyes
and ears can tell.

Suppose that your watch keeps time very
accurately—say to within a few millionths of a second per week, or better. If you set your watch from
a location near WWV in Colorado, and then hop on a
plane for Australia, you will find, when you get
there, that your watch apparently does not meet its
claimed specifications for accuracy. You might tune
in WWV on your portable shortwave radio, and
check the time setting. Strange; the tone seems to
start just a little bit too late. Not much, to be sure,
but just enough to be noticeable. But, you shouldn't
notice a millionth of a second. You check again. Still
late.

Of course, the reason for the discrepancy is
that it takes the radio waves from WWV, at the
speed of light, about 1/15 of a second to get from
Colorado to Australia. Your watch and WWV are
both very accurate. But, they do not agree in Australia, while they do in Colorado. Which instrument
is wrong in Australia? They can't both be right. Are
they perhaps both wrong?

The answer to this apparent paradox lies in the

fact that simultaneity depends on physical location. To define two events as being simultaneous, we must specify three locations or positions: those of both events, and that of an observer. In the above example, one event occurs in Colorado, and the other event and the observer are in Australia. Formerly, all three were in Colorado.

You might wonder which time source, the watch or WWV, is really correct in Australia. Astronomically, your watch is the correct reference in Australia. A time station in Australia would agree with your watch and not with WWV. There are several very accurate time stations around the world; an Australian station and WWV would agree only at observation points located equally distant from both stations. On the earth, the two stations would seem to agree everywhere on a circle running slantwise around the globe. If two time stations were set up on either pole, they would seem to agree only on the equator, for example.

In actual fact, there are time stations in Colorado, Hawaii, Japan, the USSR, and other places around the world. With this many stations, there is *no place* on the earth's surface where they all agree. Yet, they all claim to be almost exact—to within billionths of trillionths of a second! They are astronomically "synchronized," and yet there isn't a single place on earth where you would find them so by observation! You would find discrepancies of hundredths, not billionths, of a second.

CLOCKS IN SPACE

The moon is about 400,000 kilometers (km) from the earth. At 300,000 km per second, light takes approximately 1.25 seconds to go from the earth to the moon. A round trip takes 2.5 seconds. Of course, this is true of radio waves and all other electromagnetic radiation also. Perhaps you remember listening to the conversations of the Apollo astronauts with NASA. Did you notice strange gaps in the conversation? These delays were the result of the 2.5-second round-trip lag between the earth and the moon. After Ground Control finished a sentence, the astronauts would hear it 1.25 seconds later. If they replied (what would seem to them) immediately, it would be another 1.25 seconds be-

fore Ground Control would hear it.

Suppose the Apollo astronauts had left a very large clock on the moon, large enough to be seen from the earth through a powerful telescope. Or suppose they had set up a time broadcasting station on the moon. Further, imagine that they set this timepiece by tuning in the signals from WWV or some other earth-based time station (depending on which part of the earth was facing the moon at the time). Of course, these timepieces would agree on the moon; but they would differ on the earth by 2.5 seconds. They would disagree by intermediate amounts in the space between the planets. The closer to the moon an observer was located, the more nearly the clocks would agree.

On the earth, the moon-based clock (call it clock M) would be 2.5 seconds behind the earth-based clock (clock E). What about other points in space? Figure 2-1 shows this situation.

At a point midway between the earth and the moon, clock M would seem to be behind clock E by 1.25 seconds. At all points along a line connecting the earth and the moon, but on the side opposite the moon (such as point P4), the clocks would appear to be 2.5 seconds apart, just as on the earth. At all points where the moon eclipses the earth, such as P5, the clocks would seem to agree, just as on the moon. At a point in space such as P6, the clock readings would depend on the relative distances of P6 to Earth and P6 to Moon. But, at no point would the clocks disagree by more than 2.5 seconds. At no point would clock M ever seem to be ahead of clock E.

Now, suppose we changed the clock settings, so they become in agreement as viewed from some point exactly midway between the earth and the moon. This is illustrated by Fig. 2-2. The clocks would seem to agree anyplace where the distances to the two planets are the same; this set of points is a plane.

Now, on the earth, or at any point where the earth eclipses the moon, clock E will be ahead of clock M by 1.25 seconds. On the moon, or anywhere the moon eclipses the earth, clock E will be behind clock M by 1.25 seconds. At other points in space, the clocks will differ by intermediate

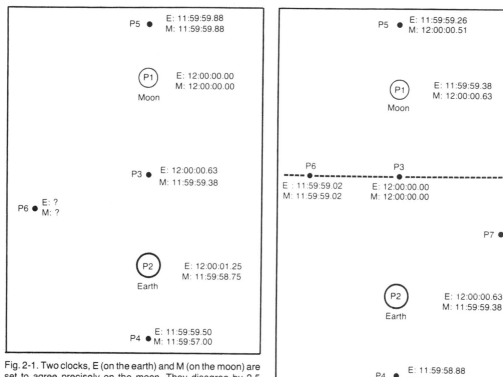

Fig. 2-1. Two clocks, E (on the earth) and M (on the moon) are set to agree precisely on the moon. They disagree by 2.5 seconds on the earth (P2), and by 1.25 seconds at the midway point between the two planets (P3). At P4, they differ by the same amount as on earth, although their actual readings are slightly behind those on earth. At P5, they agree, but are a little behind the readings on the moon. At P6, their readings will depend on the relative distances to earth and moon.

Fig. 2-2. Here, clocks E and M are set to agree at the midway point, P3, between the earth and the moon. Their reading are as shown. The readings will appear to agree at any point on a plane equidistant from the earth and the moon; in this illustration the plane cuts through the page at the dotted line.

amounts, but never will the readings differ by more than 1.25 seconds.

All this may seem a bit complicated and "messy." Isn't there any way we can get rid of this kind of time discrepancy? The answer is, unfortunately, no. The speed of electromagnetic radiation is the fastest possible speed; no means of detecting the clock readings, except by observing them or listening to time broadcasts, exists. The more clocks we attempt to synchronize, the worse this problem gets. It will take sophisticated computers to synchronize events when man becomes an interplanetary traveler!

Suppose that we have three clocks: one on Earth, one on Venus, and one on Mars. We may set these clocks to agree on Earth, but then on Venus or Mars their readings will all differ by minutes. Furthermore, as the planets revolve around the sun at their respective speeds, the difference between the readings will change. Before long, they'll have to be readjusted to agree on the earth again! How can we say what time it is when we have such a state of flux?

Perhaps there is still hope. What if we set the clocks on Venus, Earth and Mars so they agree at a point or points equidistant from all three planets? Such a point might move around as the planets revolve around the sun, but at least there will be such a point at all times. Figure 2-3 shows this arrangement.

Since the planets all follow orbits that lie almost in the same plane, known as the plane of the ecliptic, the clocks may agree at many points above and below the ecliptic as well as a point within the plane (Fig. 2-3B). When an observer is far enough above or below the plane of the ecliptic, the clocks will seem to agree at all times; or, at least, they will be very close to perfect synchronization.

Do you think we've finally come up with a solution to the problem of time synchronization? Well, we have not. Suppose we scatter clocks all over the place: on Venus, Earth, Mars, the Moon, and a few asteroids and comets that follow lopsided orbits not within the plane of the ecliptic. Suppose we launch a few satellites, with clocks aboard, around the sun in weird slanted and retrograde orbits. If there are four or more of these, in general

there will be *no* point consistently equidistant from them all. This is a simple result of 3-dimensional geometry. In this case we cannot synchronize the clocks in any convenient way. It is an utterly hopeless task; we have all these clocks, accurate to within billionths of a second, and no way to tell what time it is. Simultaneity has no meaning in 3-dimensional interplanetary space. Just like up or down or north or south, simultaneity does not exist in space.

Actually, the discrepancy of time among different points exists over very small distances as well as large distances. Instead of seconds or minutes, the discrepancy over small distances is too small to notice—trillionths of a second or less. But, in fact, *precise* simultaneity does not exist, never has and never will, unless the whole universe should collapse down to a single point. But who would know it then?

VERY FAR AGO

It's a cool night as you walk out on the dock at your little lake cottage. Cool and clear. The stars are out and the mosquitoes aren't. You lie down flat on your back on the dock, and gaze into the sky. The moon is new, so its light does not interfere with star viewing.

Some of those stars are relatively nearby, if you can call distances of hundreds of trillions of miles "nearby." The closest star, besides the sun, is so far away that light takes over four years to get to us after it has burst forth from the fiery surface. Some of the stars you see are hundreds or thousands of light years from your eyes! There's one; its light left its surface, replete with sunspots and prominences of a different sort, perhaps, from those on our sun, when Napoleon was fleeing from Moscow in the Russian winter. Or did it? Did we not say simultaneity loses its meaning in space? Then how can we say with certainty that two events—Napoleon's retreat and the departure of the photons from some star—took place at the same time? Because we're seeing the past right now! As far as we can tell, those photons leave when we see them. What other means do we have to detect the light leaving the star, until we see it? As we look at

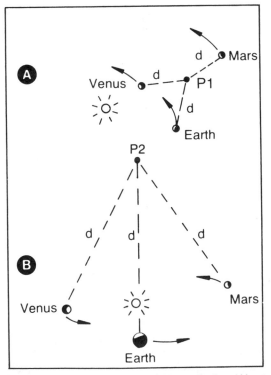

Fig. 2-3. At A, P1 is equidistant from Venus, Earth and Mars, and so the clocks on the three planets agree at P1. At B, P2 is one of many points above and below the ecliptic equidistant from the three planets, and so the clocks agree at P2.

24

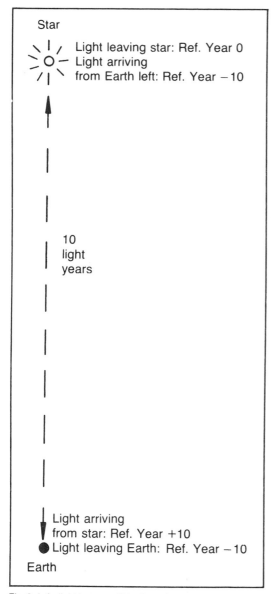

Star

Light leaving star: Ref. Year 0
Light arriving
from Earth left: Ref. Year – 10

10
light
years

Light arriving
from star: Ref. Year +10
Light leaving Earth: Ref. Year – 10
Earth

Fig. 2-4. As light leaves a distant star, images are just arriving from earth. These images are 10 years old; it will be 10 more years before the light from the star reaches earth.

that star, it is the early nineteenth century at this very moment.

You gaze at another, closer star, just 10 light years away. Light impacts on your retinas now. Suppose you could have been on that star, or near it, as those same photons were just leaving it? If you could look through an immensely powerful telescope back to our earth, what would you see? You would see the earth as it was 20 years ago. Figure 2-4 illustrates why. Light left the star 10 years ago, or 10 years before we would see it, according to some distant observer; but at that time, the light arriving at the star from earth would already be 10 years old.

Your eyes are getting well adjusted to the darkness now, and more detail is becoming visible in the sky. Suddenly, all the lights in your cottage, and all those in other houses around the lake, wink out. A power failure! How could you be so lucky? It is now totally dark except for the canopy of stars overhead. Your eyes will be able to reach their maximum degree of sensitivity. The Milky Way, our galaxy, slants across the black dome. Off to the side is a barely perceptible fuzz ball. It is so dim that you have to look slightly away from it in order to be certain it is there. Is it a small cloud? You pull out your field glasses from their case, and take a "closer" look. It has a somewhat oval shape. It is the Great Nebula in Andromeda, a different galaxy from our own. It is too far away to see any individual stars through your low-powered field glasses. The Great Nebula in Andromeda is two million light years from us.

You were alive 10 years ago. Napoleon's defeat took place a long time in the past, but only a few times your lifespan. These time periods are comprehensible. But two million years is not. Light was leaving the Andromeda galaxy at a time when no historical records were kept; a distant observer would see the photons leave that galaxy at a time when man looked considerably different, or perhaps did not even yet exist. That observer, to be sure, would not see our bustling civilization of today. This kind of thing nudges at the limits of the imagination. It is almost tempting to believe that our astronomers are wrong, that that galaxy is really just a small object within the solar system, maybe a few light *hours* away.

The discrepancy of simultaneity is great in

intergalactic space. Some galaxies are hundreds of millions, or billions, of light years distant. The difference in time between some galaxies is greater than the lifespan of the average star. The farthest galaxies we can see are several billion light years away from our own Milky Way.

THE EFFECT OF MOTION

Spatial separation affects the concept of simultaneity; so does relative motion. The effects of motion can occur as a form of Doppler effect, or as relativistic time compression. We will examine time compression in Chapter 3. For now we will look only at the Doppler effect of motion on time perception.

The Doppler effect of motion on simultaneity arises from the same cause as the Doppler effect on frequency. Suppose we have a communications link set up with a spacecraft, and the ship is traveling away from us at 300 kilometers per second (km/sec). Imagine that this ship has the ubiquitous time standard on board, ticking away the seconds with relentless accuracy. Since this ship is far away, its time readings will not agree with those of our earth-based standards. However, the disagreement will be changing constantly; it will be increasing because the ship is getting farther and farther away from us. This will cause the seconds to appear "stretched out" as we listen to the broadcasts from the space vessel.

At a speed of 300 km/sec, the ship will move 300,000 km—one light second—farther away every 1,000 seconds. We will count 1,000 ticks of our earth-based time stations during that period, but only 999 ticks from the clock on the spacecraft. The 1,000th tick will have traveled 300,000 km as we hear the 999th one, and will be passing the point where the ship was when we heard the first tick. The seconds will be 1,000/999 as long, according to the ship's clock, as they are on earth. This is approximately a tenth of one percent longer.

From the viewpoint of an astronaut aboard the space ship, the situation will seem exactly reversed. To him, the seconds from the earth-based time standards will seem 0.1 percent longer. He will hear his clock tick 1,000 times as the earth clocks tick 999 times. These situations are illustrated by Fig. 2-5 A and B.

You may wonder how this can be; how can time be "stretched" one way from one viewpoint and the opposite way from another viewpoint? Remember, though, that the ship is in motion away from the earth, and that the discrepancy in time is always getting larger. After the ship has gone 300,000 km, the discrepancy in simultaneity is one full second greater than it was before, both from the viewpoint of earthbound observers, and from the vantage point of the space traveler.

Imagine now that the astronaut completes his trip, say an excursion around the far side of the moon, and begins his return journey. Now he is approaching us at 300 km/sec. You can probably guess what will happen: the sense of time "stretching" will be reversed. Now, we on earth will hear his clock tick 1,000 times while ours ticks only 999 times. He will measure 1,000 seconds by our clock and 999 seconds by his. At such a fantastic speed, his complete round trip from the earth to the moon and back would take a meager 2,500 seconds, or just over 41 minutes! (His trajectory error margin would surely be very small at such a speed. One tiny mistake and he might careen toward the sun, or into the everlasting cold of interstellar space.)

If we observe a space ship in motion and it is not moving directly toward or away from us, but instead is traveling a diagonal course, Doppler compression will result only from that component of the vessel's velocity that is either straight toward or away from us. Figure 2-6 shows an example of this. The ship moves on a course as shown at 300 km/sec. However, its velocity along a line straight away from us is always less than 300 km/sec. As the ship progresses along this path, the angle θ gets smaller and smaller, and its radial velocity gets closer and closer to the actual speed of the vessel. Let's examine the degree of Doppler compression at the instant shown by Figure 2-6.

In order to find the component of the ship's velocity in a direction straight away from the observation point, we need to find the value of d in a triangle as shown. The velocity component will be 300d km/sec.

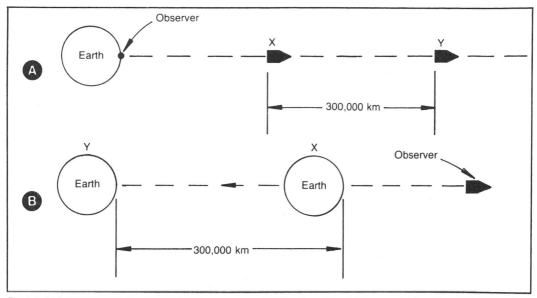

Fig. 2-5. At A, we are on earth watching a ship move away at 300 km/sec. After 1,000 seconds, the ship is 300,000 km (one light second) farther away, and signals from the vessel take one additional second to get to us. At B, we are aboard the ship; from this viewpoint, we may consider the earth to be retreating from us; signals from earth must go 300,000 km farther at time Y than at time X. The sense of time "stretching" is thus opposite for the two vantage points.

From elementary trigonometry, we can see that:

$$d/1 = \cos 30° = 0.866$$

and thus the component of the ship's velocity in line away from us is:

$$300 \times 0.866 = 260 \text{ km/sec}$$

When the ship is at point A, its velocity in line away from us is zero; it is traveling a path exactly sideways with respect to our observation point. At B, it is 260 km/sec, as we have just calculated. As the ship gets farther and farther away, and θ approaches zero degrees, the velocity in line away from us will approach 300 km/sec.

How much Doppler shift does a given speed of approach or retreat actually cause? For speeds relatively small in comparison with the speed of light c, say 0.1c or less, this is simple to determine. The speed of light is 300,000 km/sec; at a speed of 300 km/sec, we are going just 1/1,000 of the speed of

light, or 0.001c. Recall that, at 300 km/sec, the ship on its round trip of the moon caused a time "stretch" of one part in 1,000; an earthbound observer would detect 999 ticks of the spacecraft clock in a period of 1,000 seconds by his own clock. This is a Doppler compression of one part in 1,000, or 0.1 percent. In fact, the percentage of Doppler shift for speeds less than about 0.1c is approximately:

$$\text{Shift (\%)} = 100v/c,$$

where v is the speed of the object toward you, and c is the speed of light. (Speeds away from you are considered negative in the above formula.)

In Figure 2-6, at point B, the Doppler shift is thus:

$$\text{Shift (\%)} = 100 \times -260/300,000$$
$$= -0.0867\%$$

We will observe not only a change in the rate of time progression by -0.0867%, but we will also

27

notice a downward change in the frequency of the radio signal used to convey the time information. Suppose the time signal is being broadcast at a frequency of 1,000 kHz. We would actually receive it at a frequency somewhat lower than this. The frequency shift would be:

$$\text{Shift} = -0.0867\% \times 1,000 \text{ kHz}$$
$$= -0.000867 \times 1,000 \text{ kHz}$$
$$= -0.867 \text{ kHz}$$

at the observation point, to 999.133 kHz.

Let's examine the situation more closely as the ship moves from point A to point B. Suppose point A is 1,000 km from the observation point. Then, by simple trigonometry, point B is 1,733 km from point A and 2,000 km from the observation point. We will evaluate the amount of Doppler shift assuming the ship starts at point A and moves along its path at 300 km/sec, as shown in Figure 2-7. After one second, the ship has reached a point where $\theta = 73.3$ degrees:

$$\theta = \text{arcsec } (300/1,000)$$
$$= 90° - \arctan (300/1,000)$$
$$= 90° - 16.7° = 73.3°$$

Consequently its velocity component in a direction straight away from us is:

$$300 \times \cos (73.3°) \text{ km/sec}$$
$$= 300 \times 0.29 = 86.2 \text{ km/sec}$$

The percentage of shift is thus:

$$\text{Shift } (\%) = 100 \times -86.2/300,000$$
$$= -0.0287\%$$

The frequency of the signal is therefore 1,000–0.287 kHz, or 999.713 kHz.

We can calculate the frequency of the signal for any point in this same way. Figure 2-8 shows the relationship between the apparent frequency of the signal at the observation point versus the number of seconds after the ship has passed point A. The apparent frequency will approach 999 kHz, a change

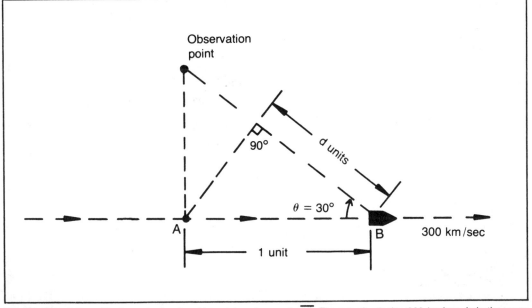

Fig. 2-6. Calculation of in-line velocity of a ship moving along path $\overline{AB}$. The actual velocity, 300 km/sec, is in the same proportion to the in-line velocity as the ratio 1 /d. (Units of distance are purely arbitrary.) The value of d is calculated by simple trigonometry.

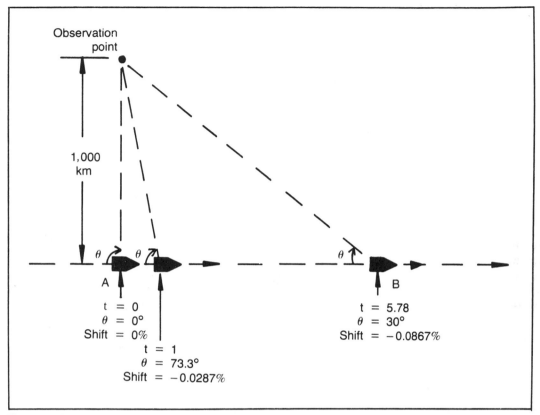

Fig. 2-7. As the ship moves along its path, passing a minimum distance of 1,000 km from the observation point, the Doppler shift increases as t increases and θ decreases. When the ship is moving in a directly lateral direction, at point A, there is no shift. At t = 1, some shift is evident. At point B (t = 5.78), the shift has nearly reached its maximum value.

of −0.1 percent, as the ship gets far away along its path.

When the speed of an object reaches about 10 percent of the speed of light, or 0.1c, another phenomenon becomes significant in the amount of frequency shift (and time compression) that we observe. We will discuss this effect, the relativistic compression of time, in Chapter 3. The preceding formulas are sufficiently accurate for speeds less than about 0.1c.

We have now seen that motion towards or away from an observer has an effect on simultaneity by causing Doppler stretching of time. We have also seen that relative position has its own effect. As a matter of fact, these two phenomena are really the very same thing: a simple result of spatial separation or a change thereof.

However, lateral motion—neither towards nor away from an observer—can also affect simultaneity. This effect, which we will now examine, is, like the others, a natural result of the constancy of the speed of light for all reference frames.

MOTION OF A VIEWER

Simultaneity can be affected not only by relative position or a change in relative position; it also depends on whether or not an observer is moving with respect to a particular set of events, and on the nature of this motion.

First, let us see what happens with sound

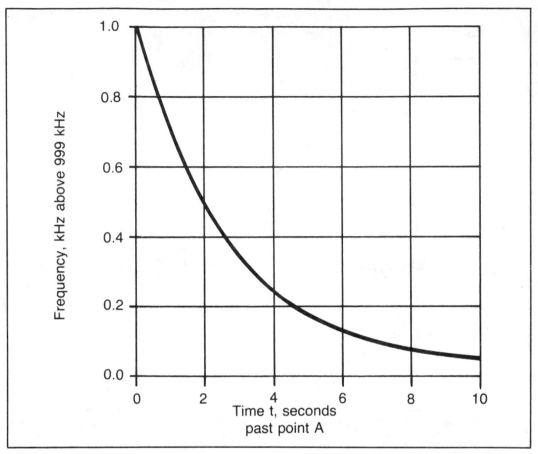

Fig. 2-8. Apparent frequency of a 1,000-kHz signal source on board the moving ship of Fig. 2-7 as a function of time. The frequency, as measured from the observation point, starts at 1,000 kHz and drops toward 999 kHz.

waves, which travel through a fixed medium, the atmosphere. Suppose we have two ocean vessels, called A and B, launched in opposite directions from some center point C at equal speed. Imagine that the two ships are launched from point C at the same time. Then the distances AC and BC are always the same, but constantly increasing. Let us also suppose that there is no wind, and that C is stationary (a buoy, perhaps, or an island). Imagine that the captains of the two vessels have set their watches to agree precisely.

Now, if someone blows a horn at point C, of course the captains of the two ships will hear it at the same moment as they look at their timepieces. The sound, traveling outward at the same speed and covering the same distance, will arrive at both ships simultaneously in the sense that the captains will agree on that fact. This is illustrated by Fig. 2-9A.

Now, if a strong, steady wind begins blowing (B), the sound, traveling at a fixed airspeed, will travel more rapidly towards one ship than towards the other. If the wind blows from C to A, the sound will get to ship A before it gets to ship B. Thus the captain of A will say the horn blew earlier. Sound moves at a definite speed with respect to a surrounding medium, no matter what the speed of the source. However, with light it is not that simple.

In the vacuum of space, there is no fixed medium in which light travels. According to the theory of relativity, the speed of light is always the same, wherever in the universe it is measured. This is true whether a point is "moving" or "still" with respect to the surrounding environment.

Let us now imagine that the ships are space vessels, and that point C is a space station situated somewhere out in the interstellar void. Suppose that C emanates a light flash rather than a noise. Prior to the departure of the ships from station C, the captains synchronize their timepieces. The ships are then launched in exactly opposite directions from C, at the same moment and at the same speed. This is shown in Fig. 2-10A.

It does not matter whether C is moving or not with respect to the background of stars in the universe. When the light flash occurs, the light will travel outward from C in all directions at equal speed; so it will overtake both A and B at the same time according to the captains of the respective vessels. If the two vessels are equipped with radio transmitters that send out bursts of signal at the instant the light flash arrives, these signals will arrive back at station C simultaneously. The signals will also arrive simultaneously at any observation station located equally distant from A and B, such as station D, assuming that station is not in motion with respect to C. As we have already seen, the signals will not arrive "in sync" at a station such as E which, while not in motion with respect to C, is not the same distance from vessels A and B.

Suppose now that this entire scenario is not

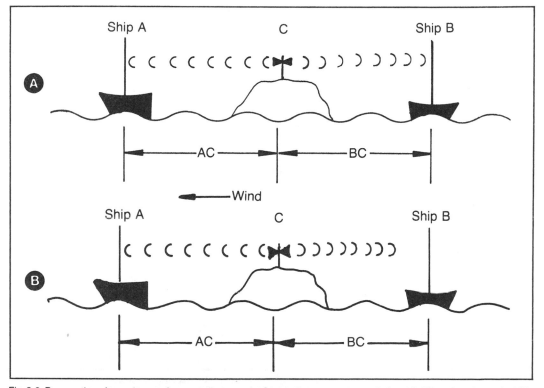

Fig. 2-9. Propagation of sound waves from a stationary point C towards two ocean vessels A and B. When there is no wind (A), the sound arrives simultaneously; when there is a wind (B), the sound gets to one ship before the other. Distances AC and BC are equal at all times.

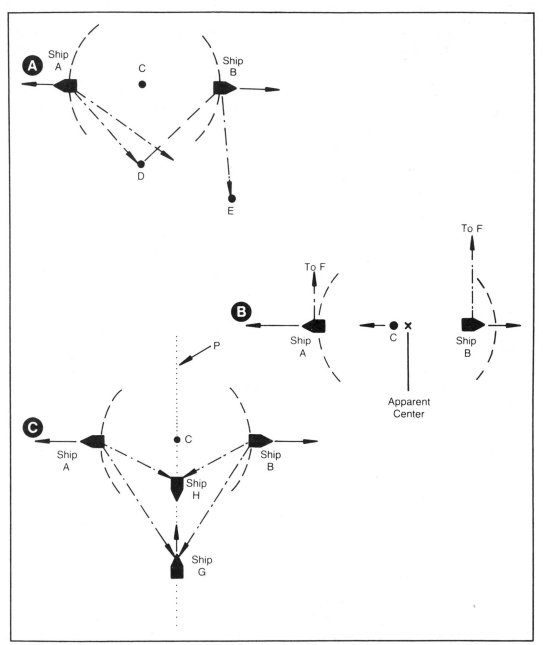

Fig. 2-10. At A, two ships are moving away from a stationary center point at equal speed, having been launched at the same time. A pulse of light sent from point C (curved dotted lines) intercepts the ships, whereupon they send radio signals (straight dotted lines). At B, the situation as observed from a distant point in lateral motion. At C, as observed from various moving ships. See text for discussion.

stationary with respect to our observation station, but is instead moving through space in the same direction ship A is traveling. This might appear to be the case from a ship F, moving in a path parallel to the path of ship B (but not necessarily at the same speed as B). Or, possibly we are on a distant planet, and C is itself in motion relative to the planet. Imagine that F is very far away from this scenario of points A, B and C—many times the distance AB—so that all three points are geometrically almost equidistant from us at all times (Fig. 2-10B). Angles ACF and BCF are almost precisely 90 degrees.

Now, the two signals sent out by ships A and B will not reach us, at F, simultaneously. Why? It cannot be because the distance AF is significantly different from BF. To illustrate the reason for this non-simultaneity, suppose we can watch the light pulse travel from C outward, maybe because of interstellar dust.

The light burst from C would appear to be a circle getting bigger and bigger around C. This circle would grow outward from a center point located where C was when the light first left C. Since C is moving, however, this center point will not be C; station C will be retreating from this point as shown in Fig. 2-10B. The circle will overtake ship B first, and then ship A. Since the radio signals travel to us from the ships at the same speed the image of the light "circle" does, we will hear the signal from B exactly as the edge of the circle crosses B. We will hear the signal from A just as the light circle overtakes A. Of course, the faster our relative speed with respect to these three points A, B and C, the greater the time difference will be.

A spacecraft G, moving on a direct line toward C perpendicular to line $\overline{ACB}$, will observe the signals from A and B at the same instant. A ship H, moving away from C along the same path, will also observe the signals simultaneously. If P represents the plane perpendicular to $\overline{ACB}$ through C, then any ship traveling in a path confined to plane P will see the signals as being "in sync." This is shown in Fig. 2-10C.

Most ships not in P will see the signals as "out of sync." It is possible that a spacecraft might be traveling in a certain path and location such that the effects of relative position and velocity just cancel each other out, making the signals from A and B appear simultaneous. From a probabilistic standpoint, this is not likely, even though an infinite number of such paths theoretically exist. We now have seen three different ways that simultaneity can be affected.

MOVING LIGHT SOURCES

There is another way to look at the influence of lateral motion on simultaneity. Perhaps you are confused by the previous example. A different illustration of this effect is given in Fig. 2-11. At A, we have an observer at a point C, midway between two hovering vessels A and B. Suppose distances AC and BC are the same, and A, B and C all lie along the same line. The operators of ships A and B send light flashes toward C, and we at point C see them at the same instant; we therefore consider the flashes to be simultaneous.

But, suppose now that we are not on the stationary outpost C, but instead on a moving vessel D, traveling along line $\overline{ACB}$ from A towards B, as shown. We will therefore pass very close to C. Imagine that observers at C, as they see the two flashes simultaneously, also see us aboard D whizzing right by. Will we, on vessel D, see both of the light flashes at the same instant we pass C? You will probably guess that the answer is no; if so, you are correct. Why?

In the relativistic sense, the situation from aboard vessel D can be thought of in either of two ways. First, we can consider A, B and C to be fixed and ourselves moving from A towards B; second, we may imagine that D is standing still and A, B and C are in motion (Fig. 2-11B). This second model is probably the better way of looking at things in order to understand why the flashes will not occur simultaneously at D.

In order for the light beams to arrive "in sync" at D just as D is passing C, the beams would have had to cover equal distances; C is exactly midway between A and B. The two beams would have had to cover equal distances from *our* point of view, which is from D. However, light doesn't travel infinitely

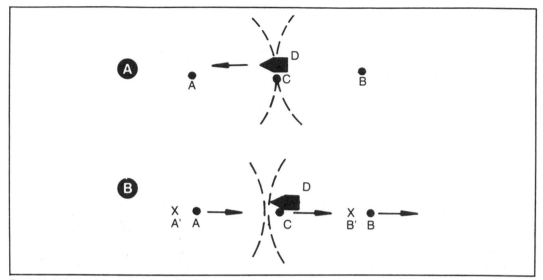

Fig. 2-11. At A, ship D is moving along a line from station A to station B, passing station C. From C, two light flashes originating at A and B arrive simultaneously. From D, they do not appear simultaneous (as shown in Fig. 2-11B); the flashes seem to emanate from A' and B', not equidistant from D.

fast. It requires some time to get from A to D and from B to D. If the flashes were to arrive at D simultaneously as D is at the midpoint, they would have had to leave their sources when A was farther away than B. Thus they would have had to travel different distances in the same amount of time. But this is not possible; the speed of light is absolute!

We must conclude, then, that the light flashes will not arrive simultaneously at D (our vantage point) as it passes C. An observer at C would think so; he would see us pass him as the flashes both occur. But we aboard D would see things differently.

RED AND BLUE SHIFT

Motion toward or away from a light source does not change the apparent speed of the light arriving from the source. That is, if we could measure the speed of a beam arriving at our space ship, that beam would appear to come in at the same speed, c, whether our motion were toward or away from the source, and regardless of our speed relative to it.

Suppose we are on board ship A as it moves away from station C in Fig. 2-10. Further suppose

that the light from C comes in the form of a continuous, monochromatic beam of light, such as a laser, perhaps in the yellow part of the visible spectrum. If our retreat from C is fast enough, this beam will seem to change color, although it will still appear to arrive at the same speed c. The color will first change to orange; speeding up the vessel sufficiently will cause it to appear red. As we speed up even more, the beam will eventually move into the infrared part of the spectrum, and become invisible. If we speed the ship up enough, we might be able to receive the beam from C on a radio! This would, of course, require extreme speed.

Now, as we slow our vessel and prepare to return to the space station C, the beam will once again become visible: first red, then orange, then yellow as we come to a halt. As we begin to accelerate toward C, the frequency of the beam will increase, moving upward through the colors of the rainbow until it fades off into the far violet. Then it will again be invisible. If we move fast enough toward C, the beam may enter the X band and become hazardous!

These effects are called "red" and "blue"

shift—decreases and increases, respectively, in frequency, caused by high speeds away from or toward a source of electromagnetic radiation. On the scale of the interstellar and intergalactic universe, such relative motion is quite common. The red and blue shifts are Doppler effects, the same effect that is responsible for the disagreement in the number of ticks counted by the earth-bound and spacecraft observers on the lunar trip discussed earlier. Red and blue shifts of light, and other electromagnetic waves, result from constantly changing spatial separation, and the resulting alteration in frequency. Each cycle of the electromagnetic field may be regarded as one "tick" of a clock.

EXPANDING SPACE

All the stars in the universe produce a certain spectrum. If we put the light of any star through a prism to spread it out into a rainbow, there will be certain spots (frequencies) in the spectrum where no light is emitted. These lines are called absorption lines, because the atoms of certain elements cause some discrete frequencies not to be radiated. This pattern of absorption lines is the same for all stars, and is easy to identify.

Sometimes this pattern of dark lines is shifted upward or downward in frequency. An upward or blue shift means that the object is moving toward us; a downward or red shift means it is moving away. Most stars and galaxies are not stationary with respect to us, and so they display some spectral shift for that component of their motion that is in a direct line with us.

The spectra of very far-off galaxies display a predominantly red shift, indicating, presumably, that they are moving away from us. The more distant a galaxy, in general, the greater the amount of red shift. This phenomenon, to be discussed in more detail in Chapter 7, has led modern astronomers to believe that the whole universe is expanding.

What does this mean with regard to simultaneity? It means that not only is there a discrepancy in time among the galaxies of the universe, amounting to millions or billions of years; the magnitude of this discrepancy is constantly changing!

Rates of time progression are being stretched by Doppler shift, and the time frame itself depends on relative location. Time, in the cosmic sense, has little meaning within our universe.

COSMIC TIME

We have seen that, because of the effects of relative position and motion on time, there cannot be a universal sense of simultaneity in our universe. Is there any possible way around this uncomfortable situation? Perhaps, at least in theory.

You will recall that there is no place on the earth's surface where time stations in Japan, Hawaii, the U.S. mainland, the USSR, and other countries will all agree. If the installations could be physically transported to the same location while still running, they would all be "in sync." But, separated as they are, no point on the planet's surface is equally distant from them all.

However, if we could go to the center of the earth, we would be at a location equidistant from all these time stations; the distance to each station would be the radius of the planet. Then we would observe all the clocks as synchronized. The trick would be to dig the hole, and to get the radio signals to travel through thousands of miles of solid and molten compressed rock and metal. Certainly, there would be practical difficulties. But, in theory, this presents a solution.

What about clocks launched into various random orbits and hovering positions in space? In our three-dimensional universe, there is nowhere we can go and be equally separated from all the clocks. In certain rare instances, this might happen, but the motion of the timepieces would soon upset things.

Imagine that we could travel into four dimensions—that there is another spatial dimension outside our three-dimensional continuum. In such a "four-space" or "hyperspace," which is admittedly impossible to visualize, we would be able to hold four straight rods together in such a way that each would be perpendicular to the other three. We will discuss the properties of hyperspace in Chapter 6; for now, let's just imagine that it exists. To facilitate visualizing the situation, consider our universe as a plane, as shown in Fig. 2-12. This is the result of "removing" one dimension, so that hyperspace can be visualized as a three-dimensional space.

While we cannot find any point in this plane where all the clocks will agree, we can get them to register pretty closely if we travel to a point in hyperspace far enough above or below the plane. On line L, passing through the plane perpendicu-larly through or near the center of the cluster of clocks, we can find points where the distances to all the clocks are nearly equal. The farther away from the plane we move along L, the more nearly equal the distances to all the clocks become, and thus the more closely they will agree.

In four dimensions, the situation is just like

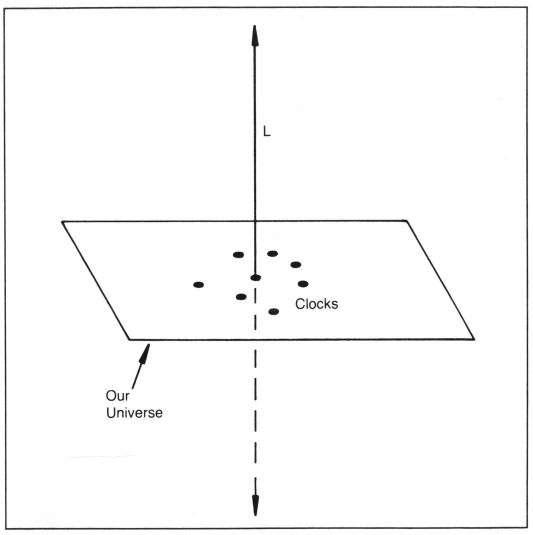

Fig. 2-12. When our universe is reduced to a plane for visualization of four-dimensional space, we can envision line L, passing through our universe at a right angle. If L passes near the center of a cluster of clocks, then points on L very far away from our universe will be almost equidistant from all the clocks.

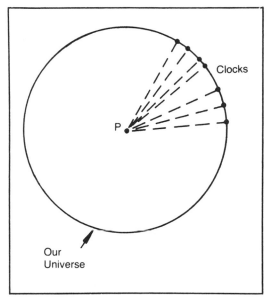

Fig. 2-13. Our universe may be a four-dimensional sphere, as shown in this dimensionally-reduced picture. From the epicenter P, all points in the universe are equidistant. Thus all the clocks, no matter where they may be, will be in perfect agreement.

line L would make a good place from which to measure cosmic time over a large area of the universe.

The universe, in the model of Fig. 2-12, appears as a plane when we remove one dimension for visualization. But it does not necessarily have to be flat. Perhaps the universe is curved with respect to hyperspace! This possibility has been pondered by cosmologists for several decades, and there is some good theoretical evidence that our universe is in fact a "hypersphere"—a set of points in four-space, all at some fixed radial distance from an epicenter P. Reduced by one dimension for visualization, the hypersphere appears as a sphere, on the surface of which we exist (Fig. 2-13).

In this illustration, it is not difficult to see that the center P of the hypersphere represents a point equidistant from every point in our universe. From such a point, all the clocks in all the strange orbits would agree—not approximately, but exactly. From this epicenter, the distance to every point on the hypersphere surface is equal to the radius of the hypersphere. Such a point, P, is the perfect place for a cosmic time reference. It is a theoretical reference only, since present technology provides us with no clues as to how we might get there.

We have seen that simultaneity has no meaning in three-space because of spatial separation and motion effects. However, extremely high speeds produce a further effect on the rate of elapsed time. This effect, known as relativistic time distortion, has intrigued and baffled astronomers and laymen ever since Einstein predicted it in the early twentieth century. Since, then, relativistic time distortion has been observed by scientists using very accurate clocks and speed measuring devices.

At speeds which are a sizable fraction of c, time goes at a slower rate. Let's find out why, and what ramifications it may have for future space travelers.

this. The trick would be sending the clock signals through four-space so we could check the readings, and getting ourselves into four-space by means of some kind of "hyperdrive." However, no matter how far out along line L we go, the clocks will always disagree just a little. If we get very far away, the disagreement caused by distance variation might be less than the inaccuracy of the clocks themselves. Still, theoretically, perfection is not possible.

There is a certain imprecision about this idea. For clock separations of millions of light years, we would have to go outward along L for fantastic distances in order to get the clocks to agree closely. Nevertheless, a point at an extreme distance on

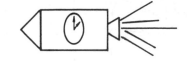

Chapter 3

Time Distortion

W E HAVE SEEN THAT THE APPARENT RATE OF time can be changed by Doppler effects when an object is moving toward or away from us. But, at sufficiently high speeds—an appreciable fraction of c—another kind of time distortion occurs. This phenomenon is known as *relativistic time distortion* or *time dilation*.

Although relativistic time distortion does not become noticeable until extreme speeds are reached, it occurs in theory at any speed, even as you walk down the street. This effect totally destroys the meaning of simultaneity, even on a cosmic scale, among objects not perfectly at rest with respect to each other.

TIME DISTORTION AS A RESULT OF MOTION

Suppose we have a spacecraft capable of attaining extreme speed, of such a nature that it can be measured in meaningful fractions of the speed of light, such as 0.3c. Imagine that we put a clock on board this vessel. When the vessel is put into space

and attains high speed, something happens to the clock inside. If this clock is synchronized with respect to an earth-based time standard at the beginning of the vessel's trip, and assuming the ship travels at high enough speed over a large enough distance, we will find that the ship's clock is running slow at the conclusion of the journey. The difference may be on the order of seconds, minutes, hours, days or years. What has happened to the clock on the ship? If we were to conduct an experiment like this without knowing about relativistic time distortion, we might conclude that the clock on the ship was malfunctioning, possibly because of the intense g force caused by the acceleration. But, any kind of time-measuring device, be it a wristwatch, wind-up alarm clock, digital electric clock, or cesium time standard, would show the same discrepancy.

Using atomic clocks capable of detecting the tiniest moments, scientists have already verified that this kind of time dilation occurs, and it takes place in the proportions that are predicted by Einstein's equations. Imagine the astonishment of scientists had they discovered this phenomenon without expecting it! Imagine the bewilderment and confusion as they tried to explain the reason for the inaccuracy of their precision instrument!

Relativistic time distortion occurs in greater and greater measure as the speed between two objects increases. If the speed is great enough, we may have a distortion factor of 2, 3, or 100. Theoretically, there is no limit to how large the factor can get; a thousand years could be compressed down into one second. The time-distortion factor increases without bound as the relative speed between two objects approaches the speed of light. At the speed of light, according to the equations, time comes to a complete halt. The speed of light is the speed of time.

Present technology has not attained the so-phistication needed to fly a space ship at speeds near the speed of light, but it may someday be possible. Then we will have to make allowances, not only for Doppler time distortion, but for relativistic effects as well.

A LIGHT-BEAM CLOCK: SPECIAL RELATIVITY

To demonstrate the reason for relativistic time distortion, we may imagine a simple sort of clock mechanism based on the constancy of the speed of light. Imagine, as illustrated by Fig. 3-1, a space ship with two mirrors placed laterally on opposite walls so that they exactly face one another. If we somehow introduce a light beam in between these two mirrors so that it strikes at a perfect right angle, we can imagine that the beam will careen back and forth between the two mirrors forever. Of course, there are practical difficulties with this idea; there is no such thing as a perfect reflector,

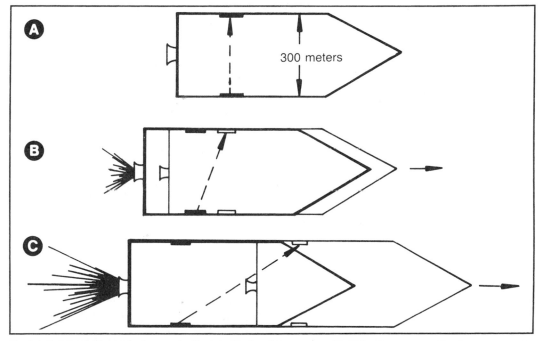

Fig. 3-1. Relativistic time distortion may be illustrated by imagining a space ship 300 meters across, with mirrors on opposite walls. A burst of light bounces back and forth between the mirrors (A); but when the ship is in motion, the light must take a slightly diagonal path because the ship moves forward as the beam moves across (B). If the speed is very great (C), the beam must travel a path considerably longer than when the ship is at rest.

and the slightest deviation from perfect orientation of the mirrors would sooner or later guide the beam off to one side, so that, in a few microseconds, it would strike the bare wall instead of the mirror. But, let us imagine perfect reflectors that are exactly parallel.

Each time the burst of light impacts against one or the other mirror, we can consider our clock to have "ticked" once. If the space ship is 300 meters wide, so that the mirrors are spaced at that distance, the "ticks" will occur at intervals of a millionth of a second, or one microsecond. (This is a pretty big space ship!)

Suppose, now, that we set this vessel in motion (Fig. 3-1B). If its speed is just a small fraction of c, the light beam will travel from one mirror to the other almost instantaneously. However, the ship will have moved just a bit in this tiny moment. In order to bounce back and forth between the two mirrors, the beam must travel a path that is slightly diagonal, or "zig-zagged." Thus, the beam will require a bit more than one microsecond to go across, according to an outside, stationary observer. From the viewpoint of a passenger aboard the ship, the burst of light will travel directly across, just as it does when the ship is at rest, over a distance of 300 meters. Therefore, while an external observer sees an increase in the time required for the beam to travel across, a passenger will not.

If the vessel is accelerated to a speed that represents a sizable fraction of c, the beam will appear to travel quite a lot more than 300 meters in order to cross the ship (Fig. 3-1C) according to an outside, stationary observer. As the speed of the ship gets close to c, the beam will appear to require more and more distance to get across. But, to a passenger on the ship, the situation will never change; it will always appear just the same as it did when the ship was at rest.

If the ship could somehow attain the velocity of light, the beam would not be able to get across the ship at all, as seen from a stationary reference point. It would have to travel faster than c to do so, and this is not possible. Inside a vessel moving at speed c, the beam of light could do no more than travel right along parallel to the walls of the ship,

according to a stationary observer.

It turns out that we can never get an object to travel at the speed of light; we shall see why in the next chapter. To do such a thing would require an infinite amount of energy. But, we are free to imagine what would happen if we *could* attain speeds of this sort: Time inside the vessel would literally stop!

POINT OF VIEW

We have mentioned the word "stationary" several times. What does this term actually mean? In the preceding discussions, we might assume "stationary" to mean "on the earth," and then assume that the vessel was in motion with respect to our planet. But we learned long ago that the earth is not the center of the universe.

Actually, defining the term "stationary" is not at all an easy task. Perhaps we can define it as a non-accelerating point. Maybe an object is "stationary" if it is not accelerating. But then, how do we explain that two non-accelerating points can be moving with respect to each other?

Maybe *any* non-accelerating point can be considered as a valid reference frame according to relativity theory. According to Newtonian physics, an object is accelerating whenever an outside force is acting on it. This rules out just about every object in the universe. The earth is being acted upon by the gravitation of the sun; the sun is being pulled toward the center of our galaxy; our galaxy is itself rotating around a center point in a cluster of galaxies, and is therefore affected by the gravity of other galaxies. And, our "local cluster" is being tugged at by other clusters of galaxies! There is no object in the universe that is not affected by some gravitational field, and thus, there is no unaccelerated object.

Perhaps we ought to define the earth as stationary, and let that be the basis for consideration of all events! But this is not very convenient in deep space. Besides, such an attitude is medieval in our time.

Actually, we can define motion only in relative terms. We will look at acceleration later; for now, we can only say that a vessel is moving at a certain velocity with respect to some other object, or with respect to the earth, or with respect to the sun. "Stationary" is a relative, not an absolute, property. In the examples shown by Fig. 3-1, we consider the outside observer to be "stationary." But it could just as well be the space ship that is fixed, and the observer that is in motion.

Imagine, then, that we have two ships, as shown in Fig. 3-2, moving with respect to each other in deep space. Suppose that both ships are equipped with mirror light-beam clocks, such as we have already seen, and that both ships are identical and are 300 meters across. Of course, according to either passenger, the clock on his own vessel will seem to be "ticking" at one-microsecond intervals. But, to either passenger, the other ship will appear to be whizzing by. Either point of view is as good as the other, and thus either observer will perceive the "clock" on the other vessel as "ticking" more slowly than his own. The greater the relative speed between the two space ships, the larger the time rate difference will be.

Relativistic time distortion or "stretching" oc-

curs regardless of the direction of the motion. In Fig. 3-3, we have two different situations; at A, two ships are passing each other on parallel courses but in opposite directions, and at B, the ships are moving directly toward each other. In both cases, the relative speed between the ships is the same, and consequently the magnitudes of the relativistic time-distortion factors are the same. Doppler effects will still occur, of course, in addition to relativistic distortion. In the example at A, there will be a blue shift first and then a red shift as the vessels pass. This will cause first a decrease, and then an increase, in the observed time distortion. At B, the blue shift will lessen the effect of relativistic distortion, and in fact will override it, making the rate of the time seem to increase. The Doppler effect in these examples makes the situation complicated; however, while Doppler shifts are the result of changes in spatial separation, the relativistic effect is an actual reduction in the rate of time.

Figure 3-4 shows an arrangement where no

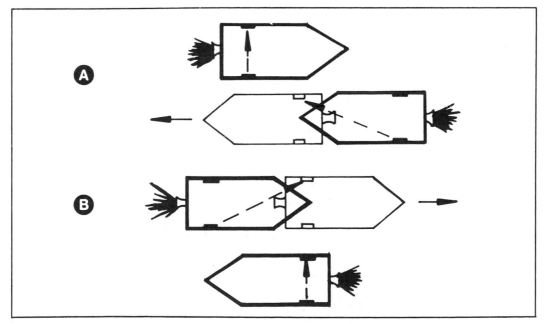

Fig. 3-2. Two different representations of two space vessels traveling along parallel courses, but in opposite directions. At A, we take the point of view of the top ship; at B, we see the situation from the reference frame of the bottom ship.

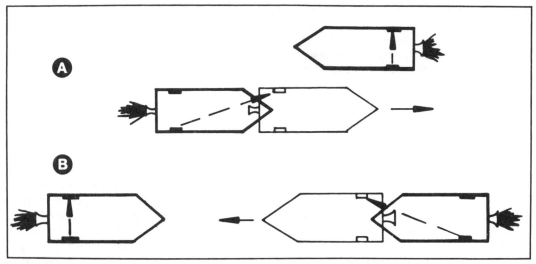

Fig. 3-3. Relative, or observed, speed may occur in any direction. At A, we take the point of view of the top ship as the vessels travel along parallel paths in opposite directions. At B, we take the point of view of the ship on the left, as the vessels move on a collision course. The amount of time distortion depends only on the observed velocity.

Doppler shifts take place, allowing us to observe only the relativistic distortion. One ship simply travels in a circle around the other. An observer in the center vessel sees the orbiting ship moving with a certain tangential velocity, and this movement causes relativistic time distortion. But, the distance between the vessels remains constant.

If the orbiting ship were started from some point in its circular path, accelerated to a high speed, and then brought back to a stop at its original starting position, we would find that the clocks aboard the two vessels were in greater disagreement than at the start. If the clocks were synchronized from the point of view of the center vessel at the start of the acceleration, they would disagree at the conclusion of the acceleration. The longer the orbiting ship was in motion, or the faster its speed, the greater the difference in the clock readings. This is not the Doppler effect, since the radial velocity of the vessels is always zero; they never get any closer together or farther apart.

CALCULATING THE TIME-DISTORTION FACTOR

As we have seen, the light beam must travel extra distance to get across the ship when the ship moves

very fast. The faster the speed of the ship, the greater this distance becomes. Let us now calculate exactly how much extra distance the beam must cover when the speed of the ship is 0.5c—half the speed of light.

Refer to Fig. 3-5. The diameter of the ship, d, is 300 meters. The path actually taken by the burst of light as it goes across the ship is shown by x; it is this distance, in meters, that we wish to find. As the beam of light travels x meters, the ship moves forward over a distance of y meters.

Since the velocity of the vessel is 0.5c in the forward direction, we know that y = x/2, because the ship must travel exactly half as far as the light beam. Remember, c is always a constant! Notice that x, y, and d, as they have been illustrated in Fig. 3-5, lie along the sides of a right triangle. (The angle between sides y and d is 90 degrees, since the mirrors are positioned laterally on opposite sides of the ship.) Recalling the Pythagorean theorem for a right triangle, we know that:

$$y^2 + d^2 = x^2$$

But, since y = x/2, we can reduce this equa-

42

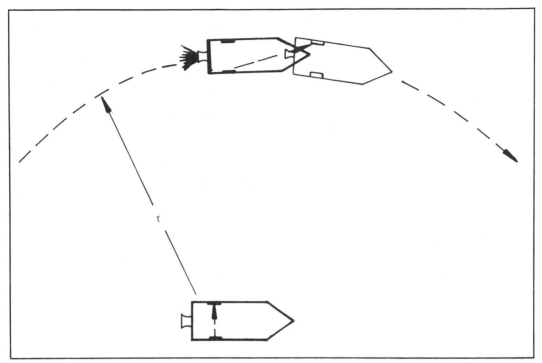

Fig. 3-4. The top vessel in this diagram moves in a circle of very large radius r around the bottom vessel. Since the two ships are always the same distance from each other, we observe no Doppler effect, but we still see relativistic time distortion.

tion to one variable, d being a constant value of 300 meters:

$$\frac{x^2}{4} + 90{,}000 = x^2$$

Solving, we first subtract $x^2/4$ from each side, giving:

$$90{,}000 = \frac{3x^2}{4},$$

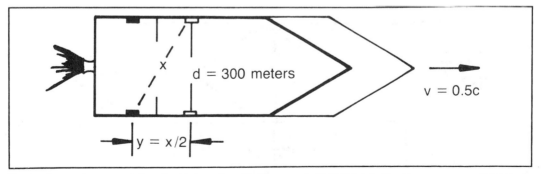

Fig. 3-5. Geometric representation of a ship in motion at half the speed of light (0.5c). As the ship moves over a distance of y meters, the light beam travels from one mirror to the other, over a distance of x meters. The diameter of the ship, d, is 300 meters.

Multiplying through by 4/3 and taking the square root gives:

$$\sqrt{120,000} = x = 346$$

The beam of light must travel 46 meters farther to get across the ship when the ship moves at 0.5c. The light therefore takes 346/300 as long to go from one side of the vessel to the other, according to an outside, stationary observer. This observer may be in another ship, on a planet, or floating around in a space suit.

Since the speed of light is a constant, the light beam will seem to require 346/300 microsecond, or 1.15 microsecond, to get across the ship as seen from the external, "stationary" point of view. But to a passenger on the ship, the situation is just the same as it is when the ship is not moving, or when it moves with any constant velocity; the passenger will always see the beam go straight across, and require just one microsecond to do it. The speed of light will appear the same to him as it does to external observers: 300,000 kilometers, or 300,000,000 meters, per second.

THE GENERAL CASE

While we have looked at one example for a specific forward speed (0.5c), and the derivation of the time-distortion factor is not very complicated, it is convenient to solve for a general velocity v, and a general fraction v/c of the speed of light. Figure 3-6 shows the geometry for this derivation. We will take the diameter of the vessel to be d, a constant; it does not matter what the actual size of the ship is. The path taken by the light burst is again given by x, and the forward progress of the ship is again given by y. Let the velocity in the forward direction be v.

The ratio y/x will be the same as the ratio v/c. The beam of light always travels at speed c, of course, and the distances are proportional to the constant speeds. If the ship is not moving with respect to the observation point, then v/c = 0, and y = 0, making y/x = 0. If the ship moves with an observed velocity of, say, 0.5c, then v/c = 0.5 and y/x = 0.5. As the speed of the ship increases, both x and y become larger and larger without limit. At a speed very close to c, both x and y will be very large distances, and they will be almost the same, since the right triangle formed by x, y, and d is very long. The beam would seem, at such a speed, to travel across the vessel very slowly, its actual path being almost parallel to the path of the ship. As v approaches c, the ratio y/x gets closer and closer to 1; we say that:

$$\lim_{v \to c} (y/x) = 1$$

This means that y/x can be as close to 1 as we want to imagine, but it is never equal to 1, nor is it ever larger.

Of course, it becomes meaningless to speak of the ratio y/x if v = c, since apparently, both x and y

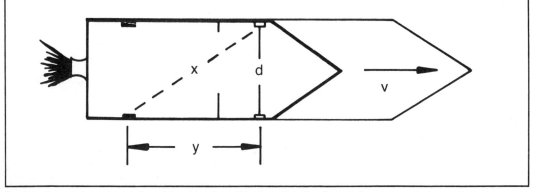

Fig. 3-6. Geometric representation of the derivation of the time-distortion formula for a generalized speed v.

become infinite at the speed of light. As we shall later see, no material object can be put in motion with speed c anyway. But, even if this were possible, the burst of light would never move away from the first mirror; the paths x and y would be identical, and the ship would continue on forever without any apparent lateral movement of the beam.

Knowing the velocity v of the vessel, we immediately know the ratio v/c, and hence the ratio y/x, since the two are equal:

$$y/x = v/c$$

Solving the above for y in terms of x gives:

$$y = x(v/c)$$

From Fig. 3-6, we can see (as in the case of Fig. 3-5) that:

$$y^2 + d^2 = x^2$$

Substituting in this equation for y in terms of x, we obtain:

$$x^2 (v^2/c^2) + d^2 = x^2$$

This is more conveniently solved if we rearrange it to:

$$x^2 (1 - v^2/c^2) = d^2$$

To find the time-distortion factor, we need the answer to the question, "How many times farther must the light beam travel when the ship is moving, as compared to when it is not moving?" Another way to put this is as follows: "How does the apparent distance traveled by the beam compare from the external viewpoint and the passenger's reference frame?" Mathematically, the time-distortion factor is given by x/d. The above equation is easily solved for this ratio by first dividing through by x^2, which gives:

$$d^2/x^2 = 1 - v^2/c^2$$

Then by inverting both sides we obtain:

$$x^2/d^2 = \frac{1}{1 - v^2/c^2}$$

Finally taking the square root, gives us the time-distortion factor:

$$x/d = \frac{1}{\sqrt{1 - v^2/c^2}}$$

By "plugging" various values of v into this equation, we can see that the time-distortion factor may be as small as 1 (when v is zero) and can grow larger without limit (as v approaches c). This is a function, and can be graphed to show relativistic time distortion as a function of speed. Figure 3-7 illustrates this.

If we suppose that v = c, we get zero in the denominator of the equation. From Fig. 3-7, it appears tempting to assume that the time-distortion factor becomes infinite when v = c. But mathematically, the value 1/0 has no meaning at all. The time-distortion factor is thus undefined at the speed of light.

We can, however, consider the inverse of the time-distortion factor as derived above. This is done by just reversing our question, so that we ask, "How many microseconds will we, as external ob-

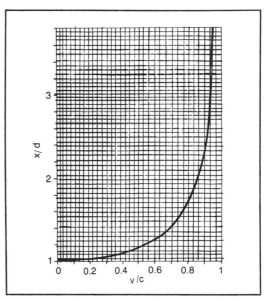

Fig. 3-7. The relativistic time-distortion factor as a function of observed velocity v in fractions of c.

servers, see go by on the ship, when our own clock registers one microsecond?" Or, "What part of the way across the ship will the light beam travel in order to cover distance d?" This ratio is d/x, and is given by:

$$d/x = \sqrt{1 - v^2/c^2}$$

This function is the reciprocal of the function illustrated by Fig. 3-7; it ranges from a value of d/x = 1 when v = 0 to d/x = 0 when v = c. Figure 3-8 shows this curve as a function of speed. Either interpretation, x/d or d/x, is equally good as far as scientific accuracy is concerned. In the next chapter, we will see that the representation x/d is better suited for expressing the mass of a moving object, and the representation d/x is better when considering the spatial length of a moving object.

"stretched" is incomprehensible. They would rather accept it as an illusion, or as the figment of a clever mathematician's imagination. However, this effect has been observed, and we cannot deny the results of experiments.

It is possible to apply the idea of time distortion to get a strange paradox, called the *Twin Paradox* or *Clock Paradox*. To resolve this paradox, we need to be a little more precise with our definitions regarding points of view. Let us look at the paradox, which at first seems quite insurmountable, and then explain the logical faults that lead up to it.

Figure 3-9 shows two vessels, launched from a space station in opposite directions at equal speed. Let us call the vessels by the names of their passengers, Mike and Joe. Both ships leave the station at the same time (simultaneity has meaning, luckily,

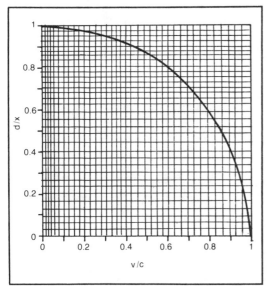

Fig. 3-8. The reciprocal of the time-distortion function in Fig. 3-7. This allows us to obtain a meaningful value for the time-distortion factor when v = c.

THE TWIN PARADOX

The preceding illustrations of relativistic time distortion are applications of the special theory of relativity. To some, the idea that time can be

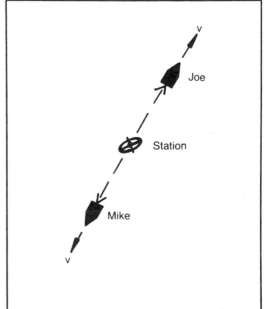

Fig. 3-9. Two space ships, piloted by Mike and Joe, are launched from the space station in opposite directions. The two ships travel "mirror-image" courses. Here, we take the point of view of the space station. When the vessels return, their clocks lag the space station clock by one minute because of relativistic time distortion. At the moment shown, both ships move with speed v.

46

when everybody is in the same place), travel at equal speeds for equal distances in opposite directions, turn around, and return, arriving back at the space station together. Mike's trip is a "mirror image" of Joe's. Suppose both Mike and Joe attain great speeds and travel for considerable distances before turning around and coming back. Imagine that Mike's clock, Joe's clock, and the station clock are all synchronized at the start of the journey, but because of the time distortion, Mike's and Joe's clocks end up a full minute behind the station clock at the end of the trip!

This sort of situation could happen in any of an infinite number of different ways. Perhaps Mike and Joe accelerate and decelerate slowly, and stay out for a long time. Maybe they accelerate and decelerate rapidly, and are gone for just a few minutes. A time discrepancy of one minute may be the result of a little distortion for quite a while, or the result of a lot of distortion for only a short while, or any in-between combination.

It should be obvious that Mike's clock and Joe's clock, while not in agreement with the clock on the station, are at least synchronized with each other. They are both one minute behind the station clock. But, we have tacitly taken the reference point of the space station for this observation. What if we take the viewpoint of Joe instead? Suddenly, things get messy.

According to Joe, the space station is in motion, and Mike is in even greater motion. This is shown by Fig. 3-10. Thus, according to the preceding derivations, the space station clock should seem to run more slowly to Joe than his own clock. In fact, since the space station appears to take the same journey (but in the opposite direction) with respect to Joe as Joe does relative to it, Joe should find the station clock a minute behind his own on the return journey! And Mike's clock will be behind by another minute—two minutes behind Joe's!

Taking the position of Mike, by exactly the same reasoning, he should see the station clock lagging his own clock by one minute, and Joe's clock lagging by two minutes! This is an utterly ridiculous situation.

How do we explain that three people together

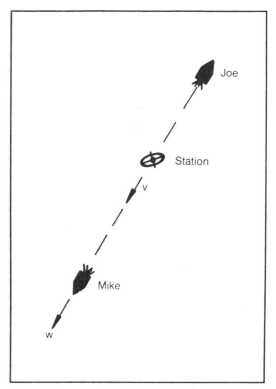

Fig. 3-10. The same situation as that of Fig. 3-9, from the vantage point of Joe. The space station moves with speed v away from Joe; Mike moves with a greater speed w, the relativistic sum of v + v.

in one room cannot agree on what their clocks say? A person eavesdropping on their conversation would think they had gone mad . . . Joe would say one thing, Mike would say another, and the scientists on board the space station would disagree with them both. Can you imagine their conversation?

This kind of contradiction is clearly unacceptable if the theory of relativity is not to be reduced to nonsense. There is another kind of paradox that arises from the same fault in reasoning in applying the special theory of relativity. This paradox is called the *Rotation Paradox*.

THE ROTATION PARADOX

Since we have essentially said that all points of view in the universe are equivalent, this should certainly

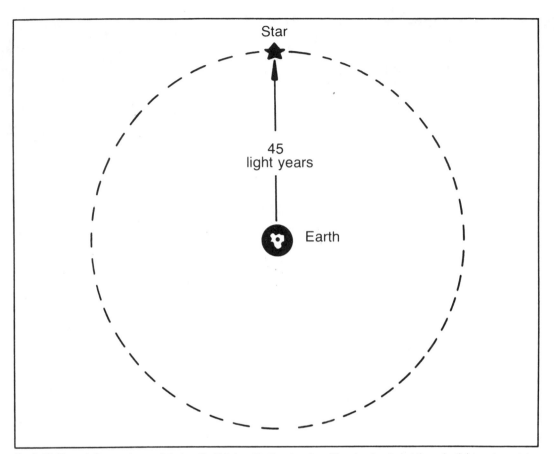

Fig. 3-11. The rotating vantage point of earth. This is a North polar view. The star, located at the celestial equator, rotates around a circle of radius 45 light years.

hold true for a rotating reference frame. In fact, our own earth is rotating. Our planet is revolving around the sun, which is in turn going around the center of the galaxy; our own Milky Way is revolving around the gravitational center of the "local group" of galaxies, and this cluster is probably revolving around some other, even more obscure, center.

From a rotating point of view, external objects are all revolving around an observer. The farther away these objects are, the greater their rotational velocity becomes. When we look at objects such as stars, their rotational speeds can be truly overwhelming!

Consider a star directly above the equator, 45 light years away from us—not very far on a galactic scale. In one day, this star travels all the way around a circle with a radius of 45 light years! This is obviously much greater than the speed of light; in a year this star moves a distance of 45 light years, times 365 days per year, times 2π or 6.28. That's 103,149 light years in a year (Fig. 3-11), or 103,149c! Most stars are much farther than 45 light years away from us, and distant galaxies—well, we can see that something is definitely wrong with this line of reasoning.

Both the Twin Paradox and the Rotation Paradox are resolved by more accurately specifying what does, and does not, constitute an unbiased reference frame in our universe. Mike's and Joe's space ships, and the rotating earth, clearly do not qualify.

STANDARDS OF MOTION

There must, then, be certain reference frames in our universe that are "better" than others. We discussed earlier that "non-accelerating" points of view might be preferable to "accelerating" ones, but then had trouble with the definition of "acceleration." We can refine this definition.

If the net gravitation on an object is zero—that is, a passenger on that object would be weightless—that object is in an *inertial* reference frame. The space station is such an object (as long as it is not one of the rotating kind designed to produce artificial gravity). To be an "unbiased" point of view, an object must be inertial. It must also be non-rotating. To ascertain rotation, we must rely on the existence of other objects in the universe, such as the distant stars and galaxies.

In the case of the Twin Paradox, Mike and Joe are on ships that are not inertial objects, since they must be accelerated and decelerated to achieve their respective speeds. This acceleration is detected by the presence of g force, which Mike and Joe will experience during periods of non-inertial motion. The rotating earth causes an outward g force (it used to be called "centrifugal force") everywhere except at its exact center, and thus our planet is not an inertial object from our vantage point on the surface. Besides the outward force caused by rotation, we are affected by a much greater inward force, called gravity; this force, too, has its effect, as we shall see in Chapter 7.

Any non-inertial reference frame can be considered as a "stationary" point for a set of events. Actually, there is no such thing as a *perfectly* inertial point of view, because of the maze of rotational vectors in our universe. But, in a localized region, we can consider such points to exist.

These two paradoxes, then, arise from false applications of the principles of special relativity. To clarify this, we explain time distortion in terms of acceleration rather than linear motion. Time distortion may be caused by the acceleration that occurs because of "pushing," such as on a space ship, and it can also happen as a result of gravitational fields.

THE GENERAL THEORY

Gravitation causes time to move more slowly. This is true of any kind of g force. The greater the intensity of the force, the greater the time distortion. We see this effect when we observe the spectral lines of stars having extremely great density. The gravitational fields at the surfaces of such stars is so great that the resulting time distortion causes a noticeable red shift. This has been observed by astronomers, lending support to the general theory of relativity.

Considering Mike and Joe once more, and their round-trip journey, it is clear that Fig. 3-9 is a valid representation of the situation, and that Fig. 3-10 is not. It is also clear that we should consider the earth to be rotating with respect to the general universe, and not the other way around, as in Fig. 3-11. Mike and Joe experience identical amounts of acceleration, and therefore their clocks will both be displaced by exactly the same amount with respect to the clock on the space station.

Gravitational fields actually cause a change in the shape of space, causing light to travel in a curved path instead of a straight line. The notion that light must always go in straight lines is no longer applicable; light always takes the shortest path it can, but if space is itself bent, it cannot possibly go straight. A more detailed look at the theory of general relativity will be given in Chapter 7. Until then, we must put off further discussion of time distortion.

ADDITION OF VELOCITIES

The special theory of relativity is useful for mathematically evaluating the addition of large velocities. For velocities small by comparison with the speed of light, we may simply add them together. Figure 3-12 illustrates this principle. Mike and Joe are both traveling away from the space station. Suppose that observers at the station see Mike retreating with velocity u, very slow compared to the speed of light, and imagine that Mike sees Joe moving away from him with speed v, also very slow by comparison with the speed of light. Since Mike and Joe are moving along the same line—that is, Mike, Joe and the station are on collinear points at all times—observers on the station will see Joe retreating at speed $w = u + v$. This is quite elementary.

But, what if both u and v are sizable fractions of the speed of light? What if $u = 0.7c$ and $v = 0.9c$? The principles of special relativity are violated if we simply add the speeds; this would give us $w = 0.7c + 0.9c = 1.6c$. But speeds in excess of the speed of light are not possible. Extremely large velocities

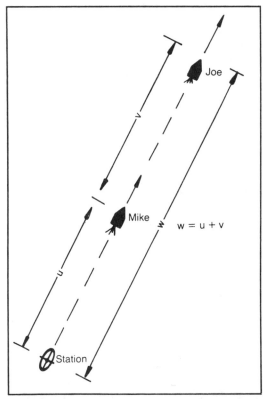

Fig. 3-12. Addition of velocities according to ordinary physics, where speeds are small compared to c. Mike moves with speed u relative to the space station, and Joe moves with speed v from the point of view of Mike. Then the space station sees Joe move with speed $w = u + v$. Mike, Joe and the station always lie on the same line in this example.

add in a different way. Actually, all velocities add according to the principles of relativity, but small velocities can be summed together for all practical purposes.

The relativistic addition of velocities can be explained by considering time distortion. Given a velocity v, recall that the relativistic time distortion is given by:

$$x/d = \frac{1}{\sqrt{1 - v^2/c^2}}$$

Suppose Mike retreats from the space station at speed $u = 0.866c$, so that the time-distortion factor, which we call x for simplicity, is equal to 2 (Fig. 3-13). Imagine that Mike sees Joe moving away from him at $v = 0.943c$, so the time-distortion factor, y, is equal to 3. (We neglect the effects of Doppler shift caused by the increasing distance between the vessels.) You can use the formula for time distortion to verify that, for velocities u and v as given, the distortion factors x and y are 2 and 3, respectively.

Station personnel will see Mike's clock move half as fast as their own, and Mike will see Joe's clock going one-third the rate of his own. Thus, it follows that the personnel in the station should see Joe's clock moving at a rate one-sixth of theirs; while they watch six seconds go by, they see Mike's clock move three seconds, and while Mike sees his three seconds elapse, he will see Joe's clock advance by one second. We are not concerned with the synchronization of the clocks, but only with their rates.

The time-dilation factor z between the station and Joe is thus 6. Using the time-distortion formula in reverse, we can calculate the speed of Joe's vessel relative to the station, as follows. Starting with:

$$z = 6 = \frac{1}{\sqrt{1 - w^2/c^2}}$$

Manipulating the variables, we obtain $w = 0.986c$ for the speed of Joe's ship.

In general, it is apparent that, for relativistic

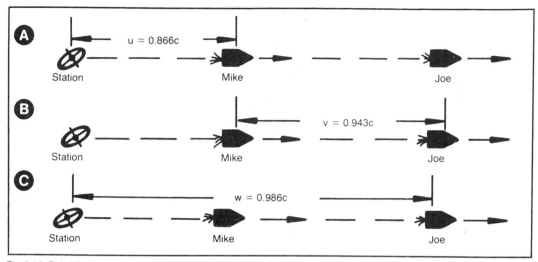

Fig. 3-13. Relativistic addition of collinear speeds. At A, observers on the station see Mike moving with speed u = 0.866c. At B, Mike oberves Joe moving with speed v = 0.943c. At C, the station sees Joe retreating with speed w = 0.986c, which is less than the impossible speed we would get if we add u and v according to ordinary physics.

addition of speeds, the time-distortion factors multiply. This is shown by Fig. 3-14. We have $z = xy$, and given the time-distortion factors:

$$x = \frac{1}{\sqrt{1 - u^2/c^2}} \qquad y = \frac{1}{\sqrt{1 - v^2/c^2}}$$

and the equation:

$$z = xy = \frac{1}{\sqrt{1 - w^2/c^2}}$$

we can calculate the relativistic sum of u and v, obtaining:

$$w = c \sqrt{1 - (1 - u^2/c^2)(1 - v^2/c^2)}$$

Now, it is apparent that the speed of light really is a maximum limiting velocity. The value under the radical in the above formula cannot exceed 1, even if $u = c$ and $v = c$. We may get extremely large values of xy; it can increase without limit. But, no matter how large xy becomes, we cannot exceed c.

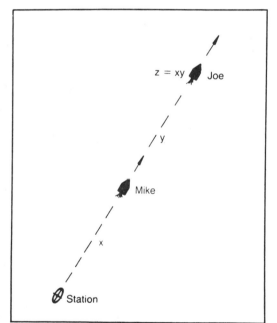

Fig. 3-14. In the relativistic addition of collinear speeds, time-distortion factors multiply. The time-distortion factor between Mike and the station is x, and the time-distortion factor between Mike and Joe is y. The time-distortion factor z between the station and Joe is thus equal to xy.

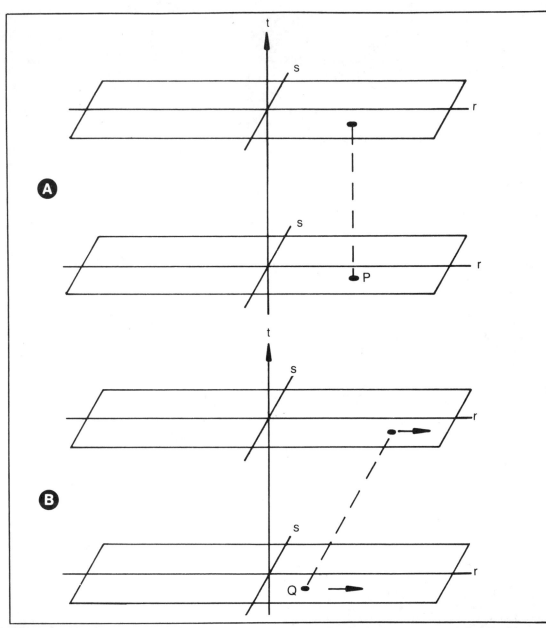

Fig. 3-15. Moving objects in space-time. At A, object P is stationary relative to the origin of the (r,s) plane. At B, object Q is in motion, thus tracing out a "world line" running diagonally through space-time. At C, R represents a photon of light. It moves across the (r,s) plane away from the origin at the same speed c as the plane moves through time; thus R follows a path that subtends a 45-degree angle with respect to the plane as well as the t axis. At D, a hypothetical object S moving through the (r,s) plane with speed greater than c. Its path subtends an angle of less than 45 degrees with respect to the universe plane.

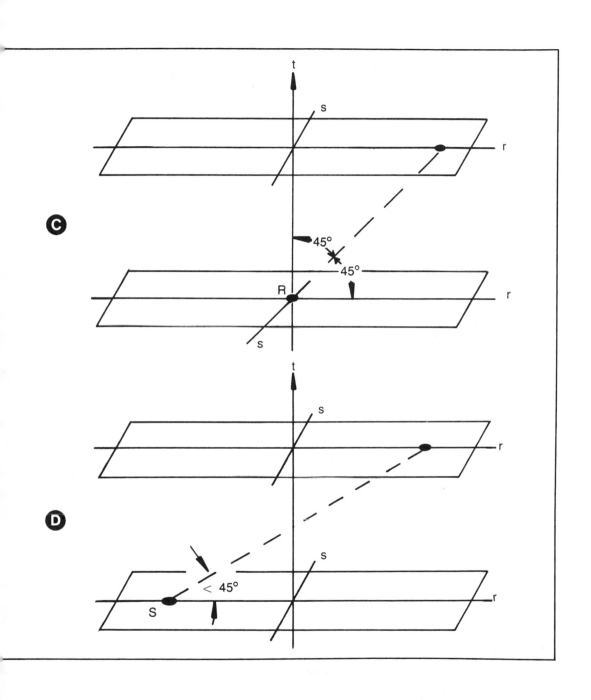

It may appear strange that the relativistic addition of c with any other speed will give c; we even have the result that c + c = c! We'll never get a chance to test this result, unfortunately, at least by means of any kind of propulsion system that we can currently imagine. But the equation leads to some interesting sorts of speculation about the nature of time and space. Perhaps there are other universes, traveling at the speed of light with respect to our own. Or perhaps faster!

TIME DIMENSIONS

The reason we cannot accelerate a material object to, or beyond, the speed of light is that its mass increases without limit as we approach c; thus an infinite amount of energy would be necessary to push a mass to that speed. We will look at the dynamics of relativity in the next chapter. Although we cannot hope to ever construct a space vessel that will travel faster than light—at least not in our own three dimensions—it is interesting to imagine other universes that might create speeds in excess of c.

Figure 3-15 is a geometric representation of the space-time continuum, with one spatial dimension removed to make the situation imaginable. The universe thus appears as a plane. The time axis, t, runs through the origin of a Cartesian (r,s) coordinate system in the universe plane. We may take any inertial point to be the origin of the (r,s) plane.

Any point that is stationary with respect to the origin, such as point P, traces out a line in the space-time continuum that is parallel to the time line t, as shown at A. An object moving at speed 0.5c, such as point Q, follows a diagonal path as illustrated at B; the universe plane is moving with speed c along the t axis from "past" to "future," and Q is moving inside the universe plane as well. A ray, or photon, of light, such as point R, emanating from the origin will follow a path that subtends a 45-degree angle to the t axis, as shown at C. Since the speed of light cannot be exceeded in the universe plane, we cannot have a "world line" at any angle smaller than 45 degrees to this plane.

But, suppose such a speed could, somehow, exist. It can be mathematically represented by a world line with an angle of less than 45 degrees to the plane. This is shown at D. As the speed becomes greater and greater, the line becomes more nearly parallel to the orientation of the universe plane. It will never become exactly parallel, however, since this would represent a speed of infinity.

Imagining that the t axis has a certain direction in a multi-dimensional time continuum, it is interesting to imagine another universe, traveling along another time axis t', which is not parallel to t. Speeds in excess of the speed of light would mathematically be possible between two such universes. One such situation is illustrated in Fig. 3-16. Objects P and Q, traveling in the two universes as nearly the speed of light, could have a relative velocity of more than c as measured in the totality of the space-time continuum.

How can time travel in different directions? All we know of time is that it seems to move from the "past" to the "future," along a time line with the "present" indicated by a continuously moving point. We never really consider seriously the possibility that time itself may have more than just one dimension, that there might be other time lines, representing universes totally separated from ours. We haven't seen any evidence of such other universes, and for good reason; Even if one of them were to collide with ours, the intersection would either be an infinitesimally thin membrane, or else would exist for just a single, mathematical time point!

Suppose we have two universes moving along time axes that are perpendicular to each other. Figure 3-17 illustrates the geometry of such a situation. Since the two time axes are mutually perpendicular, it is mathematically convenient to refer to them as the t and t i axes, where i is the imaginary number unit $\sqrt{-1}$. If each graduation on the t axis represents one second and each graduation on the t i axis represents i seconds, then we see the two universe planes moving along their axes at the rate of one graduation per second. Figure 3-17A shows an arbitrary time at which we consider the plane (r,s) to pass through t = 0 and the plane (r',s') to pass through t i = 0. After four seconds from our vantage point in "hyper-time-space," the two universe planes occupy the positions t = 4 and t i = 4 i.

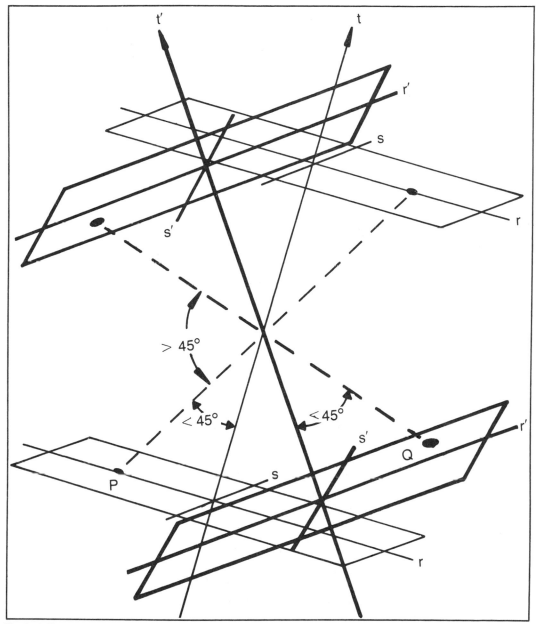

Fig. 3-16. Two universe planes (r,s) and (r',s'), moving through space-time along non-parallel time axes t and t'. Object P in plane (r,s) moves with speed less than c, and so it subtends an angle of less than 45 degrees relative to the t axis. Point Q in plane (r',s') moves with speed less than c, forming an angle smaller than 45 degrees to the t' axis. (We assume that P and Q pass through the origins of their universe planes, making such angle measurements meaningful.) However, the paths of P and Q subtend an angle of more than 45 degrees with respect to each other, and so their relative speed is greater than c.

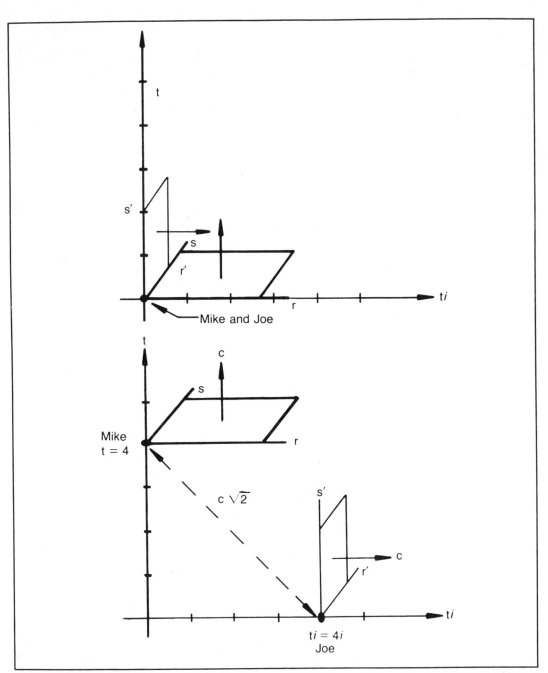

Fig. 3-17. Two universe planes (r,s) and (r′,s′) moving along perpendicular time axes. Mike and Joe are at the origins of the (r,s) and (r′,s′) planes, respectively. Their relative speed is equal to c √2, as described in the text.

Imagine that Mike exists in the universe plane (r,s), which travels along the t axis, and Joe exists in the universe plane (r',s'), which moves along the t i axis. Suppose that Mike and Joe are located at the origins of the two universe planes. Then Mike is moving with speed c along the t axis and Joe is moving with speed c along the t i axis.

By vector addition of these two speeds, which take place at right angles, we might be tempted to conclude that Mike and Joe are moving apart in "hyper-time-space" with a speed of c $\sqrt{2}$.

We can put the value c $\sqrt{2}$ into the time-distortion formula and see what happens. How many seconds t' will Joe experience relative to the viewpoint of Mike, as Mike sees one second pass? We use the formula for the reciprocal of the time-distortion factor to obtain this result:

$$t' = \sqrt{1 - v^2/c^2}$$
$$= \sqrt{1 - (c\sqrt{2})^2/c^2}$$
$$= \sqrt{1 - 2c^2/c^2} = \sqrt{1 - 2} = \sqrt{-1} = i$$

This is consistent with the model of Fig. 3-17.

While this is purely a mathematical illustration, and is based on intuitive concepts, we cannot rule out the existence of other "time dimensions." Our universe has three spatial dimensions; there is good cosmological evidence (as we will see in the last two chapters) that space may have four or more dimensions. Why could there not be several, or even infinitely many, dimensions of time as well? Even if we can't actually observe or detect universes in other time dimensions, it is interesting to imagine them—if for no other reason than simply the intuitive respite they give us, the escape they provide from the absolute limitation of the speed of light.

Chapter 4

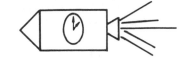

Distortion
of Space and Mass

L ARGE VELOCITIES CAUSE CHANGES IN THE SIZE and mass of moving objects, as well as in the apparent rate of time progression. When an object attains relativistic speeds, it gets heavier, and it becomes shortened or "squashed" along the axis of its motion. As the speed of an object approaches the speed of light, the mass increases without limit; mathematically we say that it "approaches infinity." The spatial length drops toward zero. These effects are observable; for example, the increase in the mass of particles in an atom smasher increases their striking force. Why do these distortions of space and mass take place? In this chapter, we will investigate models to explain these effects.

VELOCITY-CORRECTED SPACE-TIME

From the previous chapter, you will recall that we can generate dimensionally reduced models of space-time, where time is treated as a geometric dimension, along which the spatial universe travels away from the "past" and toward the "future" at the

velocity of light. Moving points trace slanted lines in this space-time model. The greater the velocity of a point in the spatial continuum, the smaller the angle created by its space-time line with respect to the continuum, and the larger the angle with respect to the time, or t, axis. In the case of photons of radiant energy, traveling with the speed of light, the angle subtended by their space-time lines is 45 degrees with respect to the spatial continuum, as well as the t axis (Fig. 3-15). Space-time lines cannot subtend an angle smaller than 45 degrees with respect to the spatial continuum, or greater than 45 degrees with respect to the t axis, because these angles would represent speeds in excess of the speed of light. That is not possible.

The space-time geometric model can still have such lines in theory; all we need to do is draw them. But such lines do not have a correspondent in reality, according to the theory of relativity. This is somewhat inelegant by mathematical standards; we do not have a one-to-one correspondence between lines in the space-time model and true velocities in the universe. We can, however, generate a space-time continuum in which such a one-to-one relation exists. Figure 4-1 shows one method of doing this.

At A, we are given an object moving with a certain velocity v, shown in the space-time continuum of Fig. 3-15 (but further reduced dimensionally, so that the spatial continuum is just a line). This object has a space-time line with a certain slope angle, say 30 degrees, relative to the t axis. The geometric slope of this line, considering the time t as the independent variable, is equal to the tangent of 30 degrees. Letting the slope be called k, then:

$$k = \tan 30° = 0.577$$

In one second, this object travels 0.577 × 300,000 km, so its velocity is 0.577c. For an object traveling at some fractional speed v/c, the slope of the line representing the object in space-time is equal to v/c, according to the model of Fig. 4-1A.

We see that the greatest possible slope for the space-time line of any object is 1. Greater slopes would represent speeds greater than that of light. But suppose we simply multiply the slope of the line by the time-distortion factor. Then, rather than v/c, we obtain a slope of:

$$k' = \frac{v/c}{\sqrt{1 - v^2/c^2}}$$

The object moving at 0.577c is then represented by a line with a geometric slope of:

$$k' = \frac{0.577}{\sqrt{1 - 0.577^2}} = 0.707$$

and hence it subtends an angle of:

$$\theta = \tan^{-1} k' = \tan^{-1} (0.707) = 35.3°$$

This is relative to the t axis. In general, the slope k' of a space-time line using this modified representation is greater than the slope k in the system depicted by Fig. 4-1A:

$$k' = \frac{k}{\sqrt{1 - k^2}}$$

For simplification, we have illustrated timelines using just one spatial dimension, but the actual situation involves three spatial dimensions. Let us refer to this modified space-time representation, such as is shown by Fig. 4-1B, as a "velocity-corrected" space-time illustration, since it (unlike the representation at A) incorporates the time-distortion factor. With this new model, any geometric line corresponds to the motion of some object. Even if the line is perpendicular to the t axis, it has a correspondent. If an object travels at the speed of light, then v = c and:

$$k' = \frac{1}{\sqrt{1 - 1^2}} = 1/0$$

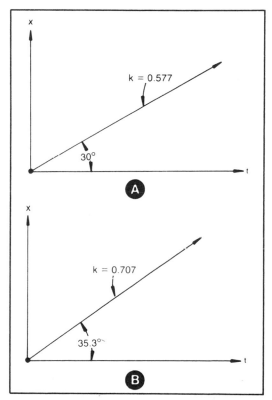

Fig. 4-1. Space-time representation of a moving object. At A, an object moving at 0.577c along a one-dimensional continuum (the x axis) traces a line with slope k = 0.577. At B, we multiply the slope by the time-distortion factor, obtaining a new slope k' = 0.707.

This means that the space-time line subtends a 90-degree angle with respect to the t axis. This is equivalent to saying that the speed of light is infinitely fast, and that a photon traverses the entire spatial continuum within one instant.

MOVING RODS

Of course, real objects are not just points; they take up a certain amount of space. In a one-dimensional spatial continuum, such as the x axis in Fig. 4-1, a real object would have some length. Motion could occur in either direction, at any speed up to the speed of light. Speeds greater than the speed of light would not be possible according to the modified space-time model, since c is represented as infinite speed in that model.

Figure 4-2 shows a rod, moving along the x axis in a space-time universe having one spatial dimension. The rod describes a band through space-time, and this band is slanted at a certain angle with respect to the t axis. The greater the speed of the rod, the greater the angle with respect to the t axis, until, at the speed of light, the rod

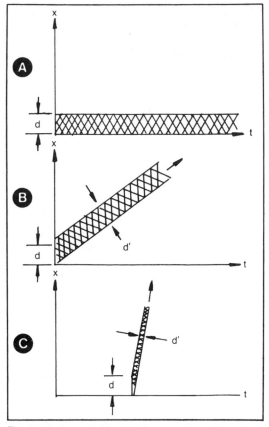

Fig. 4-3. At A, a rod of length d is at rest along the x axis. It traces out a band in space-time as shown. At B, a rod of length d moves along the x axis, tracing out a band of width d', narrower than d. At C, a rod moving in velocity-corrected space-time at almost the speed of light traces out a band of width d' much smaller than d.

describes a geometric line corresponding to the position of the x axis at one particular instant.

Suppose this rod has length d when it is at rest. When the rod is not moving, it describes a band, then, with width d, exactly parallel to the t axis (Fig. 4-3A). If the rod is moving with appreciable speed, the band becomes slanted, and its width decreases (Fig. 4-3B), although it is still the same width if measured parallel to the x axis. If the rod moves very fast, so that it generates a space-time band almost perpendicular to the t axis (Fig. 4-3C), the width of the band is much less than d. If the rod

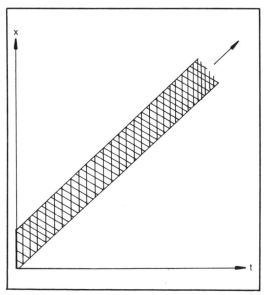

Fig. 4-2. A moving rod traces out a band in space-time. The greater the speed of the rod along the x axis, the greater the slope of the band, and the narrower the band becomes.

were to move at the speed of light, the band would become a line with zero width.

Considered in the velocity-corrected space-time universe, the rod really isn't a rod, but a band, having a width that depends on the velocity of the rod along the x axis. The apparent length of the rod is the cross-sectional measure of the band, which gets smaller and smaller as the speed of the rod gets greater.

As a specific example, suppose the rod moves along the x axis at a speed of 0.9c. Then the band consists of two parallel lines and all the region in between, both lines having a geometric slope of:

$$k' = \frac{0.9}{\sqrt{1 - 0.9^2}} = 2.1$$

This is an angle of $\theta = 64$ degrees relative to the t axis, as shown in Fig. 4-4. If the length of the rod at rest, d, is equal to 1 unit, then the width of the band, d', is given by:

$$d' = \sin (90° - \theta)$$
$$= \sin (90° - 64°) = \sin (26°) = 0.44$$

This can be seen from the geometric illustration of Figure 4-4. At 0.9c, the time-distortion factor is 2.1, and thus the width of the band in space-time is shortened by a factor equal to the time-distortion factor. Dividing the at-rest length of the rod, 1 unit, by 2.1, we obtain its in-motion apparent length of 0.44 units. This relation holds true for any speed of the rod.

In these examples, the rod can move only along its axis, and since there is only one spatial dimension, we cannot say what would happen in directions other than the direction of motion of the rod. What would happen to a rod moving in a direction not parallel to its own extension? Let us extend the model into a space-time universe of two spatial dimensions, such as is shown in Fig. 3-15. But we will use the velocity-corrected model, such that any geometric line in space-time corresponds to some point moving with uniform velocity.

MOVING DISKS

Extending the model to two spatial dimensions al-

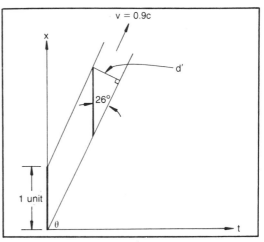

Fig. 4-4. Calculating the width d' of the band traced out by a rod of length d.

lows us to see how a disk becomes distorted when set in motion. Suppose we have a disk of diameter d, at rest in the space-time system, as shown in Fig. 4-5A. The disk describes a cylinder and all the region inside; and the center of the disk (or, for that matter, any fixed point on the disk) describes a line parallel to the t axis and perpendicular to the (x,y) plane.

Imagine now that the disk is set in motion. Its motion may or may not describe a straight line in the (x,y) plane, and this path, if it is a straight line, need not be exactly along either axis. Let's say the disk moves at uniform speed in a straight line L, passing through the origin (x,y) = (0,0) of the spatial continuum at t = 0 according to our clocks. This situation is illustrated by Fig. 4-5B.

If the disk moves rather slowly, then the space-time representation will be a cylinder that is almost perpendicular to the (x,y) plane. But, if the speed of the disk becomes very great, the cylinder will be decidedly slanted. If the disk moves with a speed almost equal to the speed of light, the cylinder will lie almost parallel to the (x,y) plane.

The greater the speed of the disk, the more flattened the cylinder will become. Although its cross section will always appear to be a disk if cut by a plane parallel to the (x,y) plane, a direct cross section will not be disk-shaped, but elliptical. In the

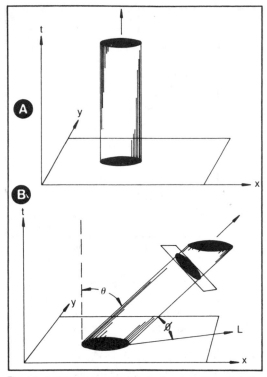

Fig. 4-5. A disk traces out a cylinder in space-time. At A, the disk is stationary, and the cylinder is circular. At B, the disk is in motion along line L, tracing out a slanted cylinder. A direct cross section of the slanted cylinder yields an ellipse, illustrating the spatial distortion caused by great speeds.

event the disk were to attain the speed of light, the disk would become a strip in the (x,y) plane at some instant, and its cross section would be a rod or line segment.

The slope of the cylinder, denoted by k', in this system is its inclination relative to the t axis. If θ represents the angular measure in degrees, then:

$$k' = \tan \theta$$

Alternatively, if ϕ is the angle measure in degrees with respect to the (x,y) plane, then:

$$k' = \tan (90° - \phi)$$

The value of k' corresponds, just as in the previous examples involving a one-dimensional

spatial universe, to the speed v/c multiplied by the time-distortion factor, or:

$$k' = \frac{v/c}{\sqrt{1 - v^2/c^2}}$$

As v approaches c, k' increases without bound, approaching a slope perpendicular to the t axis.

We can see from Fig. 4-5 that the cylinder becomes oblate, or "flattened," when the disk moves very fast. The cross section is thus not disk-shaped, but elliptical. The greater the speed of the disk, the more oblate the ellipse becomes. If the disk has diameter d at rest, then the ellipse will have a major-axis length equal to d (Fig. 4-6) regardless of the motion, but the minor-axis measure will depend on the speed.

To a stationary observer, the disk does not appear circular when it is moving with great speed. Instead, the observer sees the orthogonal cross section of the space-time cylinder. Along the direction of its motion, the disk appears shortened. At a

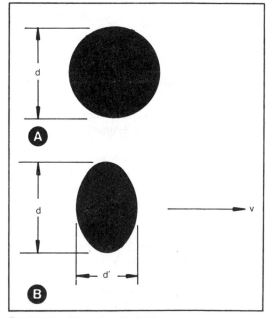

Fig. 4-6. Actual distortion of a disk in two-space. At A, the disk is at rest; at B, it is in motion with a relativistic velocity v, causing its length to appear shortened.

right angle to its direction of motion, the disk appears to have the same diameter, d, that it has at rest. When we examine an angular cross section of the disk, it seems somewhat contracted, but not as much as directly along the line of motion.

Any two-dimensional object will display this shortening effect along its axis of motion when moving at relativistically significant speeds. The factor by which the length is "compressed" is equal to the time-distortion factor. Thus, an object with a length d at rest would appear to have a length, when moving at velocity v, of:

$$d' = d \ \sqrt{1 - v^2/c^2}$$

THREE SPATIAL DIMENSIONS

The diagrams of velocity-corrected space-time become impossible to construct when the spatial continuum has three dimensions, but it is not difficult to see that the effect of high speed will be the same as it is in one or two spatial dimensions. Consider the example of a rapidly moving sphere (Fig. 4-7). The object appears to be spherical when it is at rest (A), but at relativistic speeds it becomes oblate along the axis of its motion, attaining a shape somewhat like the familiar candy-coated chocolates of our childhood days (B). As the speed of the sphere approaches the speed of light, the sphere gets so flattened that it is almost a disk (C). This effect of spatial distortion occurs only along the axis of motion, and not perpendicular to it. It is as if we were viewing moving objects through a truncated prism, or reflected in a mirror such as you may have seen in amusement centers, distorting one dimension but not the others.

MESONS

We may explain distance, or spatial, distortion in another way, as a direct corollary of the phenomenon of time distortion. This is demonstrated by particles from outer space, known as mesons.

Fast-moving protons, as well as some nuclei from various heavier atoms, enter the upper atmosphere of our planet from all directions as they arrive from the far reaches of space. As they de-

scend into the atmosphere, they collide with air molecules. When a cosmic particle hits an air molecule in the upper atmosphere, a new particle, called a meson, is produced. The lifespan of a

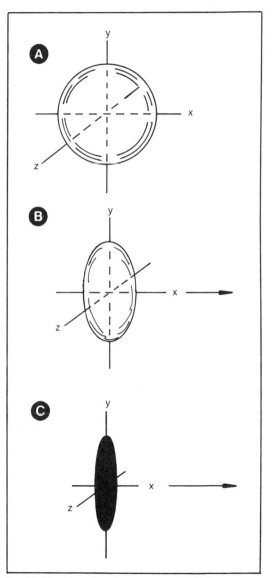

Fig. 4-7. Distortion of a sphere in space. At A, the sphere is stationary. At B, it moves at relativistic speed along the x axis, causing it to appear oblate or "flattened." At C, the speed is nearly equal to c, and the sphere is almost a disk.

meson is extremely short—they last only about 10^{-8} second—a hundredth of a microsecond.

Even if these mesons were to travel at the speed of light, they would get only about 3 meters before decaying. All mesons are generated many miles above the surface of the earth; cosmic particles cannot penetrate very far into the air before they hit a molecule. So we should think that we would never see any mesons at sea level; they should all decay far up in the atmosphere. And yet, we do see mesons on the ground! How can this be?

We may explain this apparent paradox by considering that mesons travel at almost the speed of light. In fact, they move with a speed so great that the time-distortion factor is very large. This causes the distance from the upper atmosphere to the ground to appear very small. If we could ride on a meson, the journey from the stratosphere to the ground would appear to be only a few centimeters. There would be no problem getting to sea level from high above the surface, well within the time limitation of 10^{-8} second!

This example provides us with another way to explain the apparent compression of distances at relativistic speeds. The mesons "live" for only a very short time, but as we stand on the ground and observe them, their speed is so tremendous that their lifespan is extremely prolonged because of relativistic time distortion. If we could somehow shrink ourselves to sub-atomic size and ride on a meson, we would perceive our "ship" as having a lifespan of 10^{-8} second, but this would be plenty of time to reach the ground before the meson decayed, since the distance would be greatly reduced because of the speed. From the point of view of observers on the ground, the lifespan of a meson seems to be prolonged; from the vantage point of the meson itself, we can only explain the facts by incorporating a distance-compression factor. Spatial distortion, then, takes place because *it must*, if we are to believe what we observe!

We should keep in mind that distances are only as great or small as they appear. The distance from the stratosphere to sea level does not actually change, of course, if we happen to get aboard a fast-moving vessel. We see distances from a "slant"

in space-time when moving at relativistic velocities, and from "broadside" when we are at rest. The situation is no different than the effect of watching a yardstick as it is turned to our line of vision.

Have a friend stand a hundred feet away from you, holding a yardstick exactly sideways to the line connecting himself with you. You see a certain apparent length. If he rotates the yardstick, it appears to get physically shorter, because you are looking at it from a different angle. But the yardstick is not really any shorter, although you might be led to think so by your observations. Length is as we see it.

Relativistic distortion of distance occurs in the same way, except that it is a four-dimensional space-time effect, so it is more perplexing. In the space-time universe, distances are only what we measure or perceive them to be. We can define distances only if we take a measurement!

INERTIAL MASS

When considering the relativistic effect of velocity on the mass of a moving object, it is important that we understand the distinction between inertial mass and weight. Suppose an object has an inertial mass of 28.5 grams. Then, when we put this object on a scale in the gravitational field of the earth, we will find that it has a weight of one ounce. But the same mass on Mars would weigh 0.37 ounces; on Jupiter, it would weigh about 2-½ ounces. Its inertial mass, however, would always be the same. Any object has inertial mass, and this mass is independent of the gravitational field intensity, while the weight depends on the gravitation as well as the mass. When an object is put in motion, its inertial mass increases, and this effect is what we shall now discuss.

We generally measure the weight of an object using a scale, which usually operates by means of a stretching spring, or perhaps a balance. A scale requires a gravitational field in order to function. But inertial mass can be measured in the absence of gravitation, using an oscillating-spring device such as the one shown in Fig. 4-8. The object is placed in the container between two elastic springs, mounted

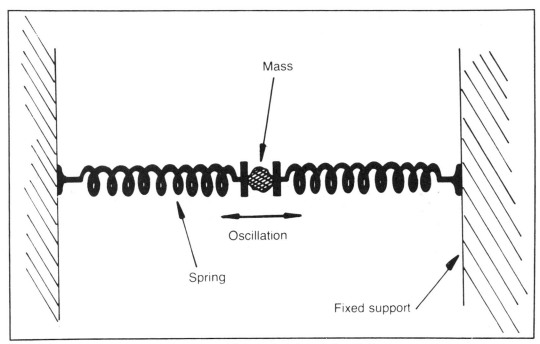

Fig. 4-8. Spring scale for determining inertial mass, independent of gravitational fields.

to fixed supports. The mass is then a function of the frequency of oscillation of the container, once it has been set in motion by pulling it to one side. The more massive the object placed in the container, the lower the frequency of oscillation will be. (You can build this kind of "mass scale" yourself, using two heavy-duty door springs.) Objects having a small inertial mass will cause the frequency of oscillation to be high, but as the inertial mass gets greater, the frequency gets progressively lower. This effect does not depend on the existence of a surrounding gravitational field, and thus it is a true measure of inertial mass. The principle of operation is, in fact, reliant on inertial resistance.

IN MOTION

Imagine that we put a spring scale of this sort on board a space ship, as shown in Fig. 4-9, and put an object in the container and set the object oscillating. Suppose also that we are always able to watch the object oscillating, so we can measure the frequency

of oscillation. When the ship is at rest with respect to us, we observe the oscillation period, and find that it is equal to one cycle per second. That is to say, it takes one second for the object to complete one trip back and forth.

Now, if we put another object in the scale, with twice the mass of the first object, the period of oscillation will double, becoming two seconds rather than one second (Fig. 4-9B). If we triple the mass of the object in the scale, the period will increase to three seconds; in general, if we put an object that is n times as massive as the original object into the scale, the period will be n seconds. Thus the period of oscillation is directly proportional to the mass in the scale. The relation is also linear; doubling the mass doubles the period of oscillation.

If the ship is set in motion, with the original mass in the scale, time distortion will cause the oscillation period to seem longer. Suppose the speed of the ship is 0.866c, such that the time-

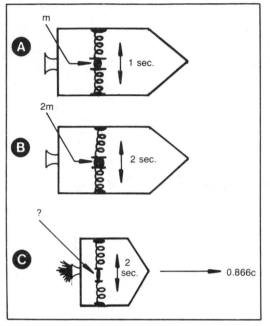

m

A   1 sec.

2m

B   2 sec.

?

C   2 sec.   0.866c

Fig. 4-9. At A, a spring scale with mass m aboard a stationary ship. The oscillation period is one second. At B, the ship is still stationary, but the mass is doubled, resulting in a period of oscillation twice as long. At C, the ship is put in motion with the original mass aboard. The oscillation period appears to have doubled. Has the mass increased to 2m? According to relativity theory, the answer is yes.

distortion factor is equal to 2 (Fig. 4-9C). Then we will observe an oscillation period of two seconds; the mass will appear to move only half as rapidly as when the vessel is at rest.

How are we to distinguish, as we watch the ship go by, between time distortion and mass increase? We cannot. We might be led to believe that the mass of the object had doubled if we did not know about time distortion. But perhaps time distortion and mass increase are actually equivalent!

If the mass of a moving object really does increase at relativistic speeds, then we should somehow be able to verify this change in mass. To watch the oscillation of a spring scale, and see that the period of oscillation increases as the speed increases, is only a theoretical observation. There must be a more definitive way to determine if the mass actually is larger.

CONSERVATION OF MOMENTUM

According to the laws of classical mechanics, the momentum p of a moving object of mass m, traveling with velocity v, is:

$$p = mv$$

One of the fundamental laws of physics states that when two bodies collide, the total momentum is the same after the collision as before. A simple example of this principle is shown in Fig. 4-10. Suppose we have two spheres of gum or wax or some other absorbent material, so that when the two bodies collide, they will remain together rather than bouncing apart. A collision wherein the objects "stick" together is called an *inelastic* collision. Suppose that both objects have identical mass m. At A, we see one sphere approaching the other with speed v; the sphere on the right appears to be standing still from our point of view. The sphere on the left therefore has momentum mv, and the stationary sphere, on the right, has zero momentum. The total momentum is thus equal to mv + 0, or simply mv.

When the two spheres collide and stick together, as shown in Fig. 4-10B, they will have a total mass of 2m. The law of conservation of momentum thus dictates that the new speed must be equal to v/2, so the total momentum can remain equal to mv:

$$p = mv = (2m) (v/2)$$

In this illustration, we have taken the vantage point of the sphere on the right. But we might equally well be situated in a position stationary to the other mass; this is shown in Fig. 4-11. In this case, the mass on the right appears to approach with velocity − v towards the mass on the left, as shown at A; when the spheres collide, the new mass is 2m and the new velocity is −v/2, so that the total momentum remains constant:

$$p = -mv = (2m) (-v/2)$$

This is shown in Fig. 4-11B. We may take any reference frame, moving in any direction at any speed with respect to either sphere, and the law of

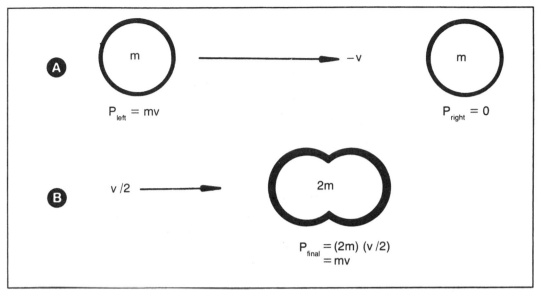

Fig. 4-10. Collision of two inelastic objects, as seen from the viewpoint of the sphere on the right. When the spheres collide, they move with a speed equal to half the original speed of the approaching mass.

conservation of momentum will always hold.

In developing a relativistic model for the dynamics of inelastic collisions—the "non-bounce" kind we have just seen—we shall require that the law of conservation of momentum never be violated. If we postulate this, then we will find that great speeds must cause a change in the mass of an object. We will now show why the notion of con-

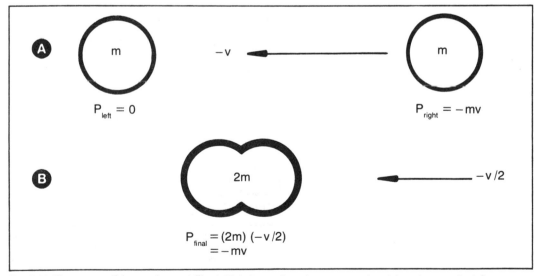

Fig. 4-11. The same collision as shown at Fig. 4-10, but seen from the vantage point of the sphere on the left.

stancy of mass must be discarded in favor of the law of conservation of momentum. When objects attain relativistic speeds, the two concepts no longer are logically consistent.

A CONTRADICTION

To show how the concept of the constancy of mass is not compatible with the law of conservation of momentum, let us say for the time being that mass is independent of velocity. Then, no matter what speed an object may attain, its inertial mass will always be equal to its rest mass.

In Fig. 4-12, we illustrate a collision between two objects moving at great speed. Let us first view the situation from the point of view of the sphere on the right. The sphere on the left approaches with great speed, say 0.9c, as shown at A. Both spheres have equal mass m, and since we have assumed that the mass is independent of the velocity, we may compute the speed of the two spheres after the collision in the following way:

$$p = 0.9cm + 0 \text{ (before)}$$
$$p = (0.45c)(2m) \text{ (after)}$$

That is, the new speed, following the inelastic collision, is 0.45c, just half the initial speed of the sphere on the left (Fig. 4-12B).

Now, let us take a viewpoint midway in between the two spheres, so that they appear to approach us from opposite directions at equal speed. Figure 4-12C shows the situation from the point of view of a space vessel, moving between the two spheres in such a way as to make it appear that both spheres are approaching from opposite directions at equal speed.

What velocity must the ship attain in order for this to be the case? If you guess 0.45c, you are wrong. Recall from Chapter 3 that if the speeds are to be equal, then the time-distortion factors must also be equal.

An observer at the sphere on the right will see a certain amount of time distortion on the ship; the observer on the ship will see the same amount of time distortion for the sphere on the left. The relative speed of the two spheres is 0.9c, which gives a time-distortion factor of:

$$x = \frac{1}{\sqrt{1 - 0.9^2}} = 2.29$$

Let us call the time-distortion factor between the ship and the sphere on the right by the name "y," and the time-distortion factor between the ship and the left-hand sphere by the name "z." Then y =

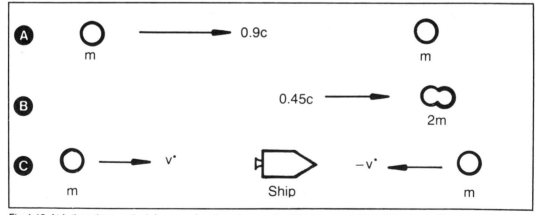

Fig. 4-12. At A, the sphere on the left approaches the sphere on the right at a speed of 0.9c. We take the viewpoint of the sphere on the right. At B, after the collision, the new speed is half the original speed of the sphere on the left, or 0.45c. At C, we take the point of view of a space ship, traveling between the spheres at such a speed as to make the spheres look like they are approaching with equal speed, but from opposite directions. This example shows how the idea of conservation of momentum implies that the mass must change with velocity.

z, because the spheres approach the ship at equal speed. Also, x = yz. Thus:

$$y = z = \sqrt{2.29} = 1.51$$

As the spheres approach the ship, the time-distortion factor is 1.51. To what velocity v* does this correspond? Working through the time-distortion formula backwards:

$$1.51 = \frac{1}{\sqrt{1 - v^{*2}/c^2}}$$

This yields a speed of v* = 0.75c. This is the speed with which an observer on the right-hand sphere will see the ship approaching, and it is also the speed with which an observer on the left-hand sphere will see himself as catching up with the ship. An observer aboard the ship will see the sphere on the left coming toward him with velocity v*; he will see the sphere on the right moving toward him with velocity −v*.

As seen from the ship, then, the sphere on the left has momentum $p_{left} = mv^*$; the sphere on the right has momentum $p_{right} = -mv^*$. Upon colliding, the final momentum will be $p_{left} + p_{right} = mv^* - mv^* = 0$. This means that the spheres would seem to come to a complete stop at the ship.

The relative velocity of the ship, taking the viewpoint of the left-hand sphere (Fig. 4-21A) is, as we have said, 0.75c. Thus the final velocity of the spheres, following the collision, is also 0.75c. This result is at variance with the previous findings, which gave a final speed of 0.45c.

This is a contradiction. Since we have resolved that the law of conservation of momentum must hold, regardless of the speed or direction of an object or objects, we are forced to conclude that the inertial mass of the moving spheres must change.

MASS VERSUS VELOCITY

The inertial mass of a moving object increases with the velocity, and this increase is exactly in proportion to the time-distortion factor. That is, when a mass of m grams (at rest) is put in motion with velocity v, it attains a mass equal to:

$$m' = \frac{m}{\sqrt{1 - v^2/c^2}}$$

The relativistic change of mass, in contrast to changes in the apparent spatial length, has demonstrable effects. A particle moving with extreme speed has greater striking force, in proportion to its velocity, than a slow-moving object. This phenomenon is important in the use of atomic particle accelerators, where the nuclei of atoms are electrically moved with speeds near the speed of light. Because the particles are moving so fast, they increase in mass and shatter the nuclei of atoms they strike. If not for this, the accelerator would not function.

For interstellar or intergalactic space travelers, the relativistic increase of mass may prove of great significance. Even in the far reaches of space, there are a few small particles—mostly hydrogen nuclei, but also an occasional small pebble as well. Moving at a speed of 0.9999c, or 99.99 percent the speed of light, a small stone having a mass of one gram (at rest) will appear to have a mass of 71 grams if it impacts against us. Its diameter will, of course, remain unchanged. At such a speed, a small object of this mass could go right through the hull of a space ship!

The increase in mass is unbounded as the velocity approaches the speed of light. That is, as we approach c (in a space vessel for instance), our mass approaches infinity. It is therefore impossible to accelerate to, or beyond, the speed of light; the faster we go, the more our inertial mass increases, and the more difficult it becomes to accelerate.

Should we ever travel to the stars, the relativistic change in the mass of our space vessel will have to be reckoned with. How will we deal with it? What effects will relativistic speeds have on us? We do not presently have the technology to find out these things in practice, but some day perhaps we will. Journeys to other stars, and other galaxies, will have far more complications than just the changes that occur in the shape and mass of space vessels. Let us now look at some of the ramifications of space travel among the stars.

Chapter 5

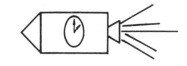

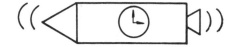

Journeys to the Stars

WHEN MEASURING DISTANCES TO THE STARS, we cannot use the familiar units of kilometers or miles; the number of kilometers or miles to any other star (besides the sun) is utterly incomprehensible. Even the sun is 150 million kilometers away from the earth, and it's hard enough to imagine this. But the nearest star outside the solar system is a staggering 27 trillion miles away, or about 44 trillion kilometers.

The Apollo space missions required about three days to go from the earth to the moon, and three days to return. Since the moon is some 400,000 kilometers from earth, this means the Apollo ships had an average speed of about 133,000 kilometers per day, or 5,500 kilometers per hour. At this speed, it would take 700 days—almost two years—to reach the sun. Consider that the distance to the Alpha Centauri star system, the closest star to our sun, is 300,000 times as far from the sun as we are! Apollo would need 600,000 years to get that far.

Look at the situation another way. If the distance between the earth and the sun were reduced to a scale of one inch, Alpha Centauri would be about 4½ miles away. The distance from the earth to the sun is called an *astronomical unit*. One astronomical unit is to a light year as an inch is to a mile. Will we *ever* reach the stars? In order to do so, we will have to attain tremendous speeds.

SPACE SHIP DESIGN FOR HIGH SPEED

If not for the effects of relativistic distortion, it is clear that we would not be able to reach many stars within the span of a human lifetime. But, by accelerating to high enough speeds, we may be able to reduce the apparent length of time required to reach other stars, and perhaps even other galaxies. The relativistic effect of extreme speed is exactly equivalent to a "slowing down" of time. The electrons in every atom, as well as the hands of any clock, move more slowly as a space vessel approaches the speed of light. The biological clock of a human being would certainly be no exception. A journey of many light years might be compressed into a time frame of only weeks, or days.

As we have seen, the inertial mass of a moving space ship increases as the speed increases, in exactly the same proportion as time becomes dilated. This means that more and more energy is required to provide additional speed. In order to take advantage of relativistic time effects, we would have to accelerate a ship to speed comparable with that of light. For a time-distortion factor of 2, we would have to attain a speed of 0.866c; for a factor of 10, 0.995c; for a time-distortion factor of 100, we would have to accelerate to 0.99995c. Even at this speed, a voyage to Alpha Centauri, 4-½ light years away, would seem to take two weeks. Traveling to stars even more distant, we would need to achieve even greater speeds. With present methods of propulsion, we could not carry enough fuel to reach this kind of velocity. The faster we go,

the more fuel we need, and the more our ship must weigh. But also, the faster we go, the greater the relativistic effect on mass becomes. It is a double-barreled problem. Until we can develop a much more efficient sort of propulsion system than the rockets we have now, we will never reach the stars.

Various design methods have already been proposed for space vessels capable of star travel. One such vessel, called "Orion," is roughly illustrated in Fig. 5-1. This ship uses nuclear bombs—hydrogen weapons—to provide acceleration. Until recently, this vessel was actually being considered for development by the United States.

A second, and smoother, way of achieving great speeds is shown in Fig. 5-2. This ship is called "Daedalus." Its propulsion system consists of a nuclear fusion reactor, a sort of controlled, continuous hydrogen bomb. Although the fusion reactor has not yet been perfected, it is likely that such a reactor will exist before too long.

Orion, designed by Theodore Taylor, Free-

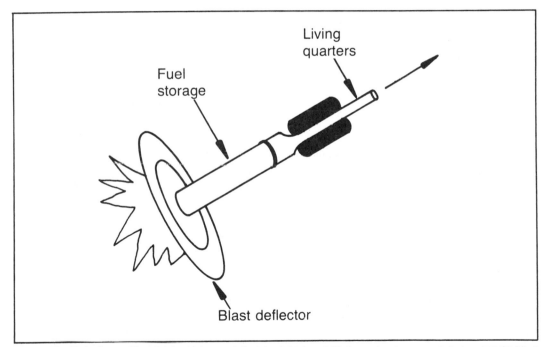

Fig. 5-1. The "Orion" vessel for attaining high speeds for interstellar travel. Hydrogen bombs are exploded against the blast deflector, pushing the ship forward. This design was proposed by Theodore Taylor, Freeman Dyson, and others.

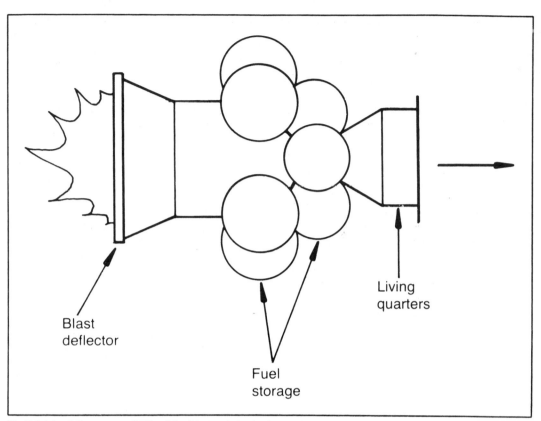

Fig. 5-2. Interstellar spacecraft "Daedalus." Instead of using bombs for propulsion, this ship uses a nuclear fusion reactor. Although such a reactor has yet to be perfected, scientists are confident that it will be some day. This ship was designed by the British Interplanetary Society.

man Dyson and other scientists, and Daedalus, designed by the British Interplanetary Society, would be capable of speeds approaching 10 percent of the speed of light—0.1c. However, at the speed the time-distortion factor is only 1.005, scarcely any different than the condition for zero velocity. Although these vessels could not take advantage of time dilation, they may be useful for rapid interplanetary travel. They are not so practical for trips to the stars; it would take over 40 years to get to Alphi Centauri, and thus it is unlikely that the same person would ever be able to say goodbye to earth and return with his story. To travel hundreds of light years would require that many generations spend their entire lives in space.

A third kind of interstellar vessel has been proposed by R. W. Bussard. This vessel may perhaps achieve relativistic speeds, since it does not have to carry its own propellant. Called the "Bussard Ramjet," this ship (Fig. 5-3) uses interstellar matter to provide acceleration, in much the same way a jet engine works in our own atmosphere. The large opening at the front gathers diffuse matter from space; it would have to be huge for the Bussard Ramjet to function. However, once it got going, this kind cf vessel could continue to accelerate indefinitely. In fact, the faster it went, the more interstellar material it would encounter, and the better it would work. Since the Bussard Ramjet does not have to carry its propellant along, it would be much less massive than Orion or Daedalus. So it would need less energy to achieve the same speed.

It is not likely that such advanced ships as Orion, Daedalus, or the Bussard Ramjet will be developed until at least several decades from now. In that time, perhaps other ideas will evolve for interstellar travel.

A TRIP TO ALPHA CENTAURI

Suppose we have a space vessel capable of accelerating indefinitely at one gravity—10 meters

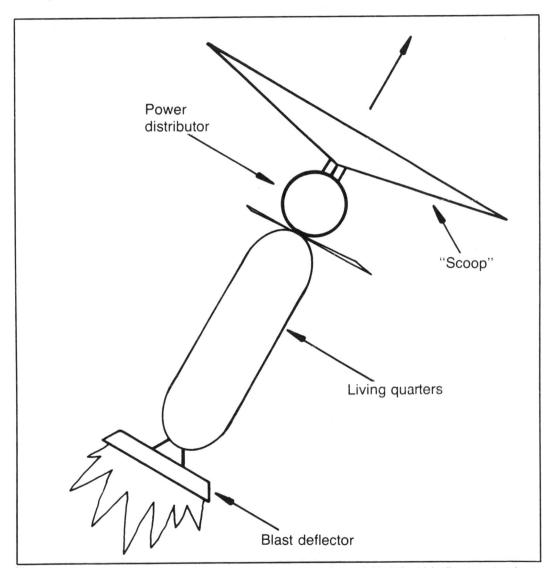

Fig. 5-3. The "Bussard Ramjet" interstellar space ship. Proposed by R. W. Bussard, this design might allow space travelers to achieve speeds great enough for relativistic time distortion to take place.

per second per second (10m/sec²)—and we embark from earth orbit in this ship, travel halfway to Alpha Centauri, turn the vessel 180 degrees, and decelerate it at one gravity until we arrive at the star. Suppose we return from Alpha Centauri to earth the same way. This will not only give us a comfortable ride, in a gravational environment familiar to us, but it will also take considerable advantage of relativistic time distortion to shorten the trip. How long would the journey appear to take us, on board the vessel? How long would our friends back home on earth have to wait for us?

Initially, our speed would be only a small fraction of the speed of light. After one minute, our speed would be $10 \, \text{m/sec}^2 \times 60 \, \text{sec} = 600 \, \text{m/sec}$, in addition to the initial orbital speed, which we will neglect since it is so small compared to the speeds we will eventually attain. After an hour, or 3,600 seconds, our speed will be $10 \, \text{m/sec}^2 \times 3,600 \, \text{sec} = 36 \, \text{km/sec}$. This is still a very small fraction of the speed of light. After a longer time, however, our speed will begin to reach values where relativistic distortion of time, mass, and distance occur. After a month, we would reach about 8.6 percent of the speed of light; after six months we will be moving at about half the speed of light. Then we must consider relativistic effects, because it is clear that time, mass, and spatial proportions will change. The vessel will eventually be speeding along at nearly the velocity of light.

Because the mass of the ship increases as it goes faster, the spatial acceleration needed to produce one gravity is less at high speeds. As the ship's speed approaches the velocity of light, its mass, and the mass of everything aboard, becomes very large. It takes much less change of velocity to create a force of one gravity when the objects are so much more massive.

As time, seen from an external point of view, goes on, the velocity of the ship levels off as it gets close to the speed of light. Figure 5-4 illustrates how this results in less and less spatial accelera-

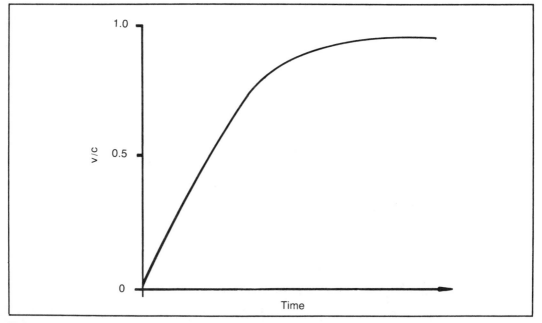

Fig. 5-4. Velocity versus time curve for a constant acceleration of one gravity. Because the space ship, and everything on it, gets more massive as the speed increases, it takes less and less spatial acceleration to create the force of one earth gravity. The result is that the speed levels off as it approaches c.

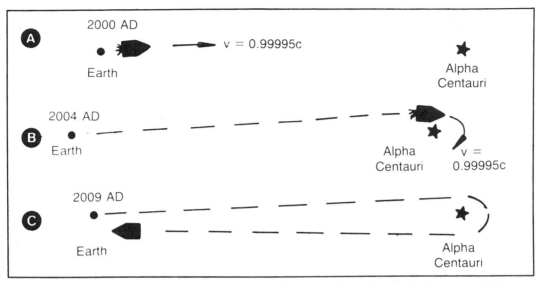

Fig. 5-5. A round-trip journey to Alpha Centauri at 99.995 percent of the speed of light. When the ship leaves earth, it is the year 2000 AD. When it returns, nine years have passed on earth, but the occupants of the vessel have aged only one month. The net result is that they have traveled into the future by eight years and 11 months.

tion. But according to the passengers on the ship, nothing changes. Time seems to them to be perfectly normal. Their weight and mass seem the same, and their surroundings appear in perfect proportion.

On a long journey, covering light years, the ship will attain speeds great enough to shorten the trip considerably for the space travelers. Instead of taking many years, reaching Alpha Centauri might seem to take only a few months.

With a space ship capable of accelerating indefinitely at one gravity, we would be able to reach great distances in space within the span of a lifetime. It would even be possible to reach other galaxies! A trip to the Andromeda galaxy, for example, would require only 28 years by ship time. But there are certain philosophical complications in all this. What father would want to see his son go off into space on a journey of hundreds of light years, never to return until long after his own death? What son would want to say goodbye to his father, knowing it would be for the last time? On a journey lasting hundreds, thousands, or millions of years, with animation suspended on the ship by a large

factor, how would space travelers feel as they left the earth? Their home would certainly be different when they returned; it might not even exist, considering the capability of mankind to destroy our planet.

It is not an important consideration for space travelers of today. But, will future astronauts ever volunteer for a long relativistic journey? Will future governments allow such expeditions? Space travel may never be easy.

TIME TRAVEL

Let us imagine a space ship capable of quickly accelerating to, and decelerating from, relativistic speeds. Such acceleration and deceleration would introduce extreme gravitational effects on the passengers. However, perhaps the technology of the future will provide a way to overcome this. Anyway, we might attain very large time-distortion factors aboard such ships—so great that time travel would become a reality. It would be possible to literally travel to the future!

Figure 5-5 shows an imaginary journey to and from Alpha Centauri, which is almost 4½ light years from our solar system. At a speed of 99.995 percent the speed of light, or 0.99995c, the time-distortion factor is 100. Imagine that this vessel makes a circumnavigation of Alpha Centauri, leaving earth in the year 2000. When it returns, it will have traveled nine light years at almost the speed of light, and the passengers will find it is the year 2009. But, they will have seen only a hundredth of that time span pass—only about a month. Their bodies will have aged just one month; their clocks will have moved ahead the equivalent of only one month. These astronauts will have moved ahead in time by nearly nine years.

The consequences of this kind of journey would be pretty strange. Suppose you are 28, and your father is 50; after such a time journey, you would still be 28, but your father would be 59. Or suppose you are 28 and have a younger sister who is 21. When you returned, she would be your older sister.

Time journeys of greater magnitude are possible; theoretically, there is no limit to how far into the future we might go. If our vessel were to accel-

erate to sufficient speed, and circumnavigate the Andromeda galaxy, a round trip of about four million light years, the time-distortion factor could be 10^6 (Fig. 5-6). On such a trip, we might get into our vessel, be cloistered for 48 months in the strange depths of space-time, and return to earth to find— well, what? Would the planet still be inhabited? Would technology have found a way to avoid relativistic effects of intergalactic travel? Would life be better or worse? Of course we would have no way of knowing when we began the trip. But, of one thing we could be certain: The trade-off would be final. There is no known return from a relativistic voyage into the future. To travel backward in time is, apparently, impossible, because it creates logical contradictions.

BACKWARD TIME TRAVEL

Looking back at the space-time models of Fig. 3-15, we can represent a relativistic journey into the future, such as the round trip to Alpha Centauri or the Andromeda galaxy, by a pictorial model (Fig.

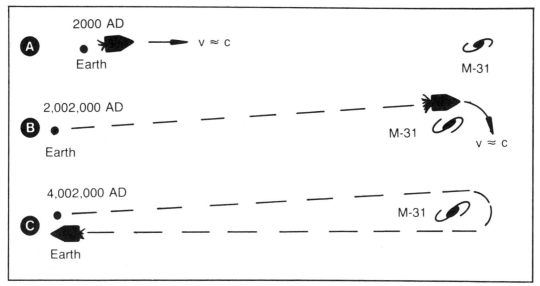

Fig. 5-6. A round-trip excursion to M-31, the Andromeda Galaxy, which is two million light years from our solar system. If the ship leaves in the year 2000 AD, it will return four million years later. However, if the vessel reaches a speed v almost the speed of light, so that the time distortion factor is 10^6 (a million), the occupants of the ship will have aged just 48 months. Theoretically, any time-distortion factor can be achieved, and so such a trip to the future is possible.

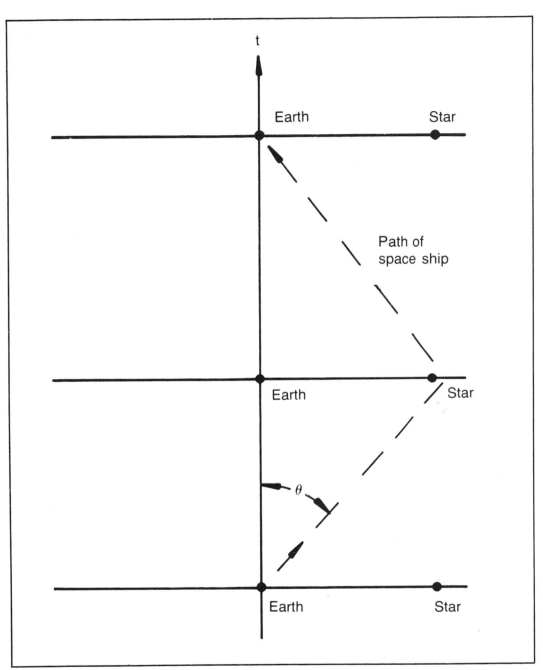

Fig. 5-7. Space-time representation of a journey such as that depicted by Fig. 5-5 or 5-6. The path of the ship is shown by the dotted line. The angle θ is almost 45 degrees; if it were 45 degrees then the speed of the ship would be c.

5-7). At near-light velocity, we follow a path inclined almost 45 degrees with respect to the time axis. Upon our return to the time line, running through our home planet earth, it is the future.

Much fiction has been written about backward time travel. It will surely remain fiction, unless and until we find ways to leave our three-dimensional universe. Relativistic travel into the future presents no contradictions; it is essentially a state of suspended animation. But traveling into the past would make it possible to change history. This would create a fragile cosmos indeed!

Imagine what might happen if we could travel into the past. Such a journey might be represented by the space-time model of Fig. 5-8. What would it be like if we could travel through the space-time continuum freely? You might go back 24 hours, and correct a mistake you made yesterday on an examination. You might go farther into the past, and change history. History is like a tree: It is a continual succession of causes and effects. The farther back you went, the greater the impact your actions could have. You might even prevent your own birth! This would keep you from making the time journey in the first place.

Clearly, this contradiction is intolerable. It reduces the whole universe to nonsense. We must dismiss backward time travel. And it is a good thing we can't do it! Any one of us might just disappear at any moment!

DIFFICULTIES WITH INTERSTELLAR TRAVEL

Relativistic effects on the rate of time may make interstellar space travel less tedious, although we have seen there could be psychological problems associated with long journeys. But, the relativistic effect on mass has a detrimental effect on the prospects for traveling among the stars.

We know that space is not a perfect vacuum; there are atoms of hydrogen here and there, even in the remotest reaches of space. The density of matter in the universe varies, but it averages perhaps one atom of hydrogen per cubic centimeter. It is this matter that allows the Bussard Ramjet to achieve its high speeds. But, at greater and greater speed, these molecules will begin to drag on a space ship, because it will encounter more and more of them per unit time. Also, at relativistic velocities, the effective mass of each atom will increase. According to physicist Edward Pucell at a lecture in 1960, if you were traveling at 0.99c, the force of the interstellar materials against your space vessel would result in radiation equivalent to that produced by several hundred particle accelerators, for each square yard of surface area! It is clear that this would present a safety hazard. And 0.99c results in a time-distortion factor of only 7. Particle accelerators, also called "atom smashers," rely partly on the relativistic effects of great speed in order to work. But this creates a lot of danger for would-be interstellar travelers. How would they shield their vessel without making it prohibitively heavy? How serious would this situation get at a speed where the time-distortion factor was 100, 1,000, or more?

The effective mass caused by relativistic speed increases not only for atoms of hydrogen, but for all interstellar material, including any meteors that might happen to lie in the ship's way. The greater the speed, the higher the chance that a meteor will be encountered. Suppose we are cruising along at 99.995 percent the speed of light, and happen to run into a marble-sized piece of interstellar iron ore with a mass of, say, an ounce. Because the distortion factor is 100 at this speed, the little rock will have an effective mass of over six pounds. Needless to say, a six-pound object, a half inch across, moving at 99.995 percent the speed of light, would have a lot of momentum. It would surely do damage to our ship!

There are other problems with relativistic space travel besides the dangers of radiation (Fig. 5-9) and meteors (Fig. 5-10). Time distortion, while making long journeys endurable or at least possible, will also cause certain inconveniences, aside from the perhaps undesirable time-travel effect. Let's look at a couple of these problems.

TELECOMMUNICATIONS

Time distortion will cause a Doppler-like decrease in the frequency of radio signals that might be used for communications between earth and a relativistic space vessel. This, in addition to actual Doppler

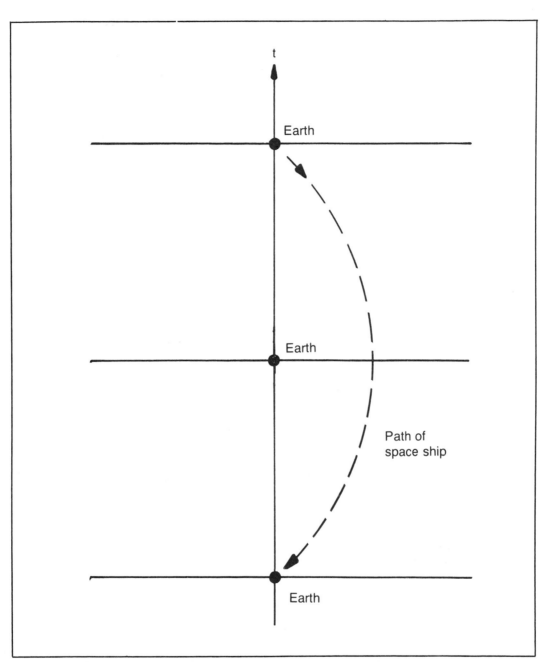

Fig. 5-8. A trip back in time, pictorially represented by a space-time drawing. Such trips are impossible, according to our present knowledge. The ship moves in the opposite direction from that of our universe, hurtling into the future at the speed of light. This ship must therefore travel outside our space continuum.

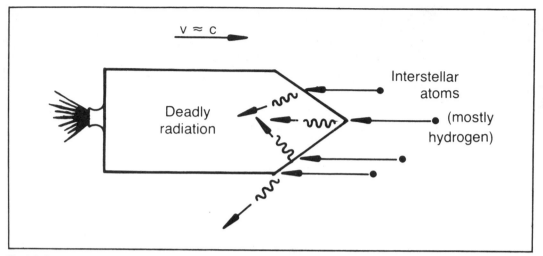

Fig. 5-9. As a space vessel moves at nearly the speed of light through interstellar space, the diffuse atoms, mostly of hydrogen, strike the front of the ship with greatly increased mass. This produces deadly radiation because of the "atom smashing" effect. In fact, according to physicist Edward Purcell, each square yard of the ship would be hit by the equivalent output of several hundred particle accelerators!

shift resulting from motion, will not be a very serious problem by itself. We will simply have to make adjustments to the frequency setting of our receiver, both aboard the ship, and back on earth. But aside from this frequency difference, time will be moving at different rates, too! A two-way conversation will be utterly out of the question. The great distances between the earth and an interstellar spacecraft will make this an impossibility; even a one-way conversation (exchange of data) will be quite prolonged, once the vessel is a few hundreds or thousands of light years away.

The time-distortion effect will stretch out the modulation of whatever signals do eventually arrive at earth from the vessel, or vice-versa. When this factor becomes large, special equipment will be necessary to decode the low frequencies. Rather than voices, it will be necessary to use a telecom-

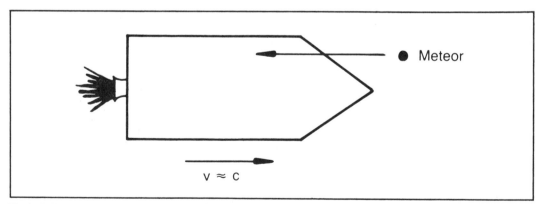

Fig. 5-10. Even the smallest meteor would become a deadly projectile at near-light velocities. Not only would it strike the ship at tremendous speed, but its mass would be relativistically increased, and consequently its density, too.

munication code such as high-speed Morse. But with a time-distortion factor of a million, even the fastest data will be rendered ridiculously slow. Information transmitted at 1,000 words per minute would be reduced to one word every 16 hours and 40 minutes. Patience will certainly be an ingredient of interstellar or intergalactic voyages!

STAR HOPPING

Science-fiction books and movies that depict future space travelers making excursions among the stars, perhaps covering many cubic light years of space, conveniently neglect the extreme time delays involved in such "star hopping."

Suppose an interstellar vessel leaves Planet X, orbiting Star X, and makes a "short" trip to Planet Y, which orbits Star Y, to obtain some necessary supplies for the people of Planet X. Figure 5-11 shows one such situation, where Star X and Star Y are ten light years away from each other. These two stars are very close together when compared to the diameter of our galaxy. (The Milky

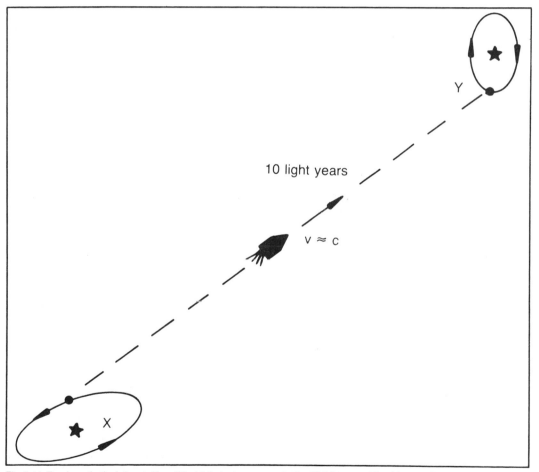

10 light years

v ≈ c

Fig. 5-11. Two hypothetical planets X and Y orbit their stars, separated by 10 light years. The ship, traveling at almost the speed of light, is on its way to Planet Y to get some crucial supplies for Planet X. It has already been gone five years. There are still 15 years left until it arrives with the supplies back at Planet X.

Way is nearly 100,000 light years across.) Let's imagine that Planets X and Y, because of their proximity, have come to identify with the same "interplanetary organization."

Even if the spacecraft accelerates to almost the velocity of light during its excursion, it will be 20 years by Planet X time until it returns with the crucial supplies. We are, of course, speaking here about time in earth years; Planet X may have a year that is longer or shorter than ours, if its orbit takes more or less time to go once around its sun. But 20

earth years is quite a while by anybody's standards. Still, two stars only ten light years apart are not likely to both have planets with civilizations. Two stars, both with sophisticated technological societies on at least one of their respective planets, would more likely be many thousands, or even millions, of light years away from each other. We can make an optimistic guess, though, and suppose that they are only about 1,500 light years apart.

Imagine what an interstellar "summit meeting" between the two great superpower planets of a

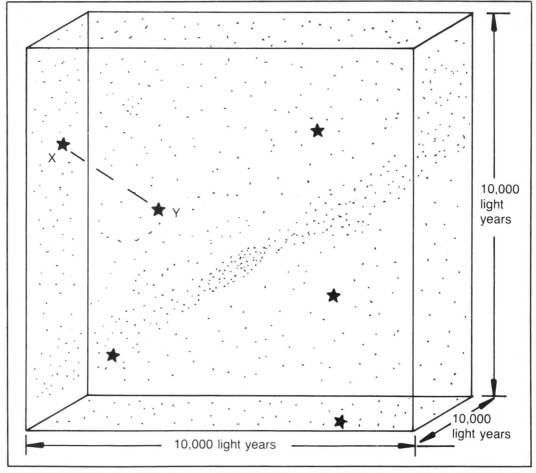

Fig. 5-12. In a cubical section of the galaxy measuring 10,000 light years on each edge, there are six stars (shown as having points) with intelligent civilizations. These six societies have formed a federation. Planets X and Y are by far the most advanced. They are 1,500 light years apart. How will they hold a summit meeting?

"federation," enclosed by a cubical chunk of space 10,000 light years on a side, would be like! Suppose the two superpowers, Planets X and Y, are situated as shown in Fig. 5-12. (It is fortunate that they are not at opposite corners of the cube! But they are still awfully far away from each other.)

The superpowers might conduct their meeting either in person or by radio or television. But either way, the process would take many generations, if these creatures have lifespans anything at all like our own. Such societies would surely have to be much different from ours. Perhaps patience evolves with technological advancement, although with us, it seems to have been the other way around.

Science fiction authors have invented "hyperlight drives" and "subspace communication" devices to get around this kind of problem, while still providing a believable scenario. But even at ten times the speed of light, assuming it were possible to go that fast, the journey from Planet X to Planet Y would require 150 years. "Subspace" communication would be the only feasible way to have an interstellar meeting. But, if two planets were separated by distances such that a trip from one to the other would span generations, we would think their civilizations would not get very involved with each other. They might communicate for entertainment, in much the same way as our amateur radio operators do today. But further contact would probably be very limited.

This is all speculation, of course. According to relativity theory, it must also be quite fictitious. But we are just beginning to make discoveries about very strange aspects of the cosmos. It might be possible to travel extreme distances in almost no time at all, in a manner that is totally unknown to us at the present time.

Hopefully, we have learned a lesson from the arrogance of those in the past. Just when we think we know everything, something new and radical is discovered. The "hyper-light drive" invented by science-fiction enthusiasts may one day be found to be fact! We should never completely give up the idea of interstellar or intergalactic travel.

HAVE WE BEEN VISITED?

With our current knowledge, we can imagine how interstellar travel might be possible, although inconvenient, dangerous and perhaps psychologically traumatic. Would other advanced civilizations ever attempt interstellar or intergalactic journeys? Is it possible that some of the apparent difficulties can be overcome? One way to find out is to search for evidence of a visitation, to our earth, by beings from another world.

We have all heard about unidentified flying objects (UFOs); some of us have even seen them. In many UFO reports, the witnesses have described extremely rapid acceleration and deceleration, such that the vessels dart about almost like projected images or holograms. Dismissing, for a moment, the fact that not a single one of these "vessels" has ever contacted our leaders, how is it possible for them to accelerate and decelerate so fast? Wouldn't the forces introduced on the passengers be overwhelming, crushing them to death? It would seem so. Maybe they have found a way to neutralize this effect. If this is true, perhaps they have conquered other difficulties with long-distance space travel as well.

Unfortunately, we cannot say with certainty whether or not UFOs actually represent visitations by extraterrestrial civilizations. Although many UFO stories lack precise explanations, we need more concrete evidence before we can reasonably and scientifically conclude that these events are the result of visits by beings from other planets. Some day they may contact us in such a way that there can be no doubt about their identity. Until then, we must look for more definite evidence.

CLOSE ENCOUNTERS

There are a few cases where people have claimed to have been physically transported away from earth by a UFO. Upon careful questioning, and even hypnosis, they have given answers that defy explanation. It is impossible to rule out a visitation by aliens; such occasions brighten our hopes for interstellar travel. But, we have still not had any aliens land at the Kennedy Space Center and show us how their ships operate.

In the Bible's Old Testament, the book of Ezekiel makes reference to strange objects floating in the sky, and winged creatures that transported the prophet over great distances. There are strange stories in other legends, as well. There have always been unexplained phenomena. Maybe some of these stories stem from visitations long ago, and maybe not. There are bizarre figures etched into the ground in one desert region of South America. From ground level, they look like ruts made by a tractor or dug in the earth by means of shovels. But when viewed from the sky, these etchings take on definite shapes; they are mostly animal-like in appearance. Were these "drawings" made to please the gods? Or are they communications signals, perhaps intended as landing instructions for alien space ships?

It would seem that, if interstellar or intergalactic travelers had come to earth and wanted us humans to be aware of their existence, they would have left some sort of unmistakable evidence, such as star maps, drawings of their vessels and themselves, or records in one of our languages. But nothing of this sort has ever been discovered. There are always several plausible explanations for the strange phenomena that we are tempted to attribute to visitors from space.

Frank Drake, a 1952 graduate of Cornell University, and other scientists, think that perhaps radiocommunication might be used, instead of space travel, as a means by which societies from other planets might try to contact us. Radio waves

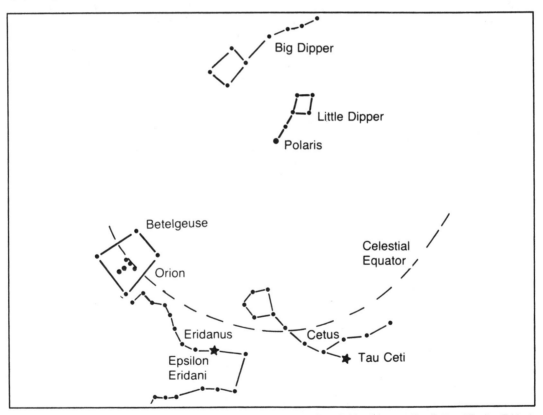

Fig. 5-13. Star map showing the approximate locations, in the sky as seen from Earth, of the stars Tau Ceti and Epsilon Eridani. The dotted line represents the celestial equator, 90 degrees from the North celestial pole (located near Polaris). These two stars were the focus of the initiation of Project Ozma, the first serious attempt to listen for signals from other worlds. These two stars are about 11 light years from earth.

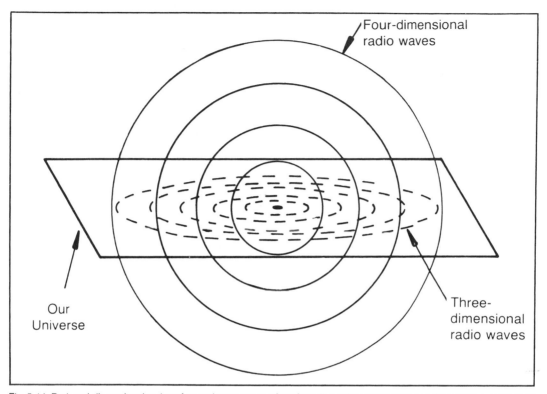

Fig. 5-14. Reduced-dimension drawing of our universe as seen from four-dimensional "hyperspace." Our continuum appears as a plane. Ordinary radio waves travel within our universe as shown by the circles in the plane. Is it possible to generate waves that travel outside our three-dimensional universe? Such electromagnetic energy would, in this drawing, appear as spheres around the emission point. (Here, they are shown as circles outside the plane.)

travel at the fastest possible speed, and no living beings would need to sacrifice their lives in a relativistic time journey if this method of contact were used. Maybe we are being sought after at this very minute by radio beacons sent from a civilization on another planet, orbiting another star!

PROJECT OZMA

Frank Drake initiated *Project Ozma* in April of 1959. Named after "the princess of the imaginary land of Oz," this was the first serious attempt to receive signals from other worlds. It was greeted with mixed reactions from scientists—some were as enthusiastic as Drake and his fellow astronomers, and some were not only skeptical, but almost scornful. Nevertheless, Drake believed that, if

enough stars were scanned near the 21-centimeter resonant hydrogen wavelength, signals from another civilization would eventually be heard. And not in a million years, either: It might actually happen on the third or fourth star that was checked!

The 21-centimeter wavelength was chosen because it, as the fundamental resonant frequency of hydrogen atoms in space, is perhaps the most common radio frequency in the cosmos. It seemed that this wavelength would be used by anyone wishing to send signals intended for reception by beings of other solar systems.

Among the stars considered for priority checking were Tau Ceti, in the constellation Cetus, and Epsilon Eridani, in the constellation Eridanus. Figure 5-13 shows the locations of these stars in the heavens as seen from our planet. Both stars are fairly close to our sun—about 11 light years away. Drake showed that signals of sufficient power to be heard by our radio telescopes could be generated by any society having mastery of the techniques of radio communication. And any reasonably advanced beings would, Drake believed, have completely mastered the electromagnetic spectrum.

The search began; at one point, strong, regular pulses were received. This caused great excitement for Drake and his colleagues. But, they finally found that the signals were actually of earthly origin, apparently transmitted by the military in a radar experiment. The astronomers reiterated their requests for radio silence in the vicinity of their sensitive equipment.

The quest for reception of signals from other planets goes on. Astronomers still look for such transmissions, since the rewards would be well worth whatever effort and cost are involved. Are they looking on the correct frequencies? Is there perhaps another way by which radio signals can be transmitted, a way that we have not yet discovered?

"HYPERSPACE" COMMUNICATION

If there were only some way to shorten the vast distances between the stars and galaxies, at least for radio transmissions, if not physical transport! There may in fact be, but it would involve the transmission of energy or matter into more than three spatial dimensions. Perhaps in a space of many dimensions, speeds faster than light are possible; it could be that radio signals might travel at a higher speed in four dimensions than they do in three dimensions. Or, possibly, distances may be reduced in "hyperspace."

It is impossible for us to directly visualize a space of four or more dimensions. But, we can geometrically describe what might take place in such a "hyperspace." Figure 5-14 is a diagram of what our three-dimensional continuum would look like in four-dimensional hyperspace; one dimension has been taken away to make it comprehensible. Four-dimensional radio signals are shown propagating outside our own universe into hyperspace. Is it possible that ordinary radio waves do this? Or would we have to invent some kind of special antenna or transmitter? Perhaps the whole idea is utterly out of the question; maybe it is completely ridiculous to suppose that any such thing is possible. But we don't know this yet!

If we could ever find a way to generate radio waves of more than three dimensions, perhaps we might communicate with other civilizations, avoiding the tremendous delays involved with interstellar transmission of ordinary electromagnetic signals. The fiction of the present may become the fact of the future.

Cosmologists have very good evidence that four-dimensional hyperspace does exist. Often, we consider time as a fourth dimension, perhaps because we cannot visualize a space of more than three. This is a logical model, but it is quite likely that there are other spatial dimensions. Let's examine the properties of hyperspaces of four or more dimensions. What would it be like to live in a four-dimensional universe?

Chapter 6

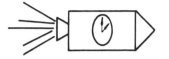

Dimensions and Hyperspace

STRONOMERS AND COSMOLOGISTS OF TODAY believe that there are more than three dimensions. We cannot visualize "hyperspaces" of four or more dimensions, but we can geometrically explore them, and get an intuitive picture of what they are like. It is quite simple to define coordinate systems, distances, and volumes in spaces of four or more dimensions.

WHAT IS A DIMENSION?

Without realizing it, perhaps, we have used the term "dimensions" quite frequently without giving it a definition. We can agree that our universe is three-dimensional, or that space-time is, in some sense, four-dimensional. A geometric plane, or the surface of a sphere, are examples of two-dimensional spaces. A line and a circle are examples of one-dimensional spaces. Perhaps we should define what it means to have a certain number of dimensions, rather than attempt to define the term "dimension" itself.

In our universe, we can construct three lines through any point, so that they all intersect there and are all perpendicular to each other. Figure 6-1 shows this. A good way to visualize it is to imagine a corner in a room, where two walls meet the ceiling. One line is formed by the intersection of the two walls, and runs vertically; the other two lines are formed by the intersection of the walls with the ceiling, and they run in a horizontal direction. They are mutually orthogonal at the point of intersection; each line is at a right angle to the other two.

In our universe, every point has the property that we can find exactly three lines that are mutually perpendicular. Never can we find four or more, and we can always find three. If we could find a point where four mutually perpendicular lines were possible, we would have a four-dimensional space, or hyperspace. If we were only able to find two such lines, we would have a two-space. An example of a two-space is the surface of our own earth. The surface of a sphere is such that, in the locality of any

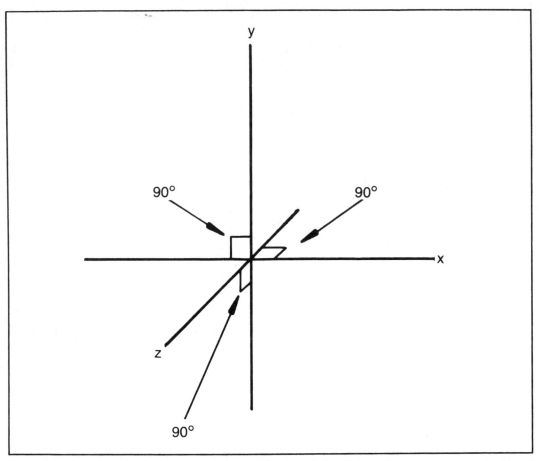

Fig. 6-1. In our three-dimensional universe, we can choose three geometric lines that pass through a common point and are all perpendicular to each other. Here, we call the lines x, y and z.

one point, only two lines can be found that are at right angles to one another (Fig. 6-2).

We may define the concept of dimensionality, then, in terms of how many mutually perpendicular lines we can find in a given universe. A space is n-dimensional if, but only if, we can find exactly n lines through any point in the space, such that all the lines are perpendicular at that point.

Two things are obvious about this. First, n will always be a whole number; we will never have a space of three and a half dimensions! Second, we will not consider the possibility that a space might have a different number of dimensions in some places than in others; if any one point has room for n mutually orthogonal lines, then we shall assume that all points in the whole universe are the same way.

Another way to define the dimensionality of a universe is to see how many coordinates are needed to locate a given point. How many coordinate values are needed, for example, to define the location of a point on the surface of the earth? Of course, the answer is two, and the most common coordinate method for finding geographic locations on earth is the latitude/longitude system. If we consider all space around the earth, then we must add a third coordinate to determine, uniquely, all points in the universe. A two-dimensional space requires two coordinates to establish the exact position of a point, and a three-dimensional space requires three (Fig. 6-3). In general, an n-dimensional space will have the property that any coordinate scheme, defining the unique location of a point, will need to make use of exactly n different numbers. Each point will then correspond to a certain combination of such values. That is, different points will always have different coordinates, and a given point will have only one set of numbers that identifies it. Either the orthogonal-line scheme or the coordinate scheme may be used to provide a definition for dimensionality.

THE CARTESIAN SYSTEM

There are various different kinds of coordinate systems, but the simplest and most straightforward (for most applications) is called the *Cartesian coor-*

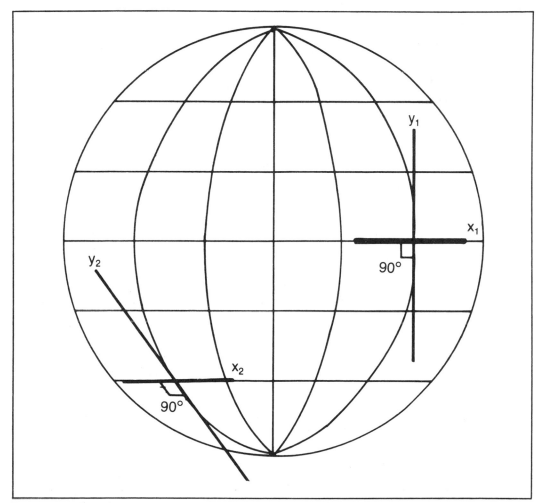

Fig. 6-2. The surface of a sphere is two-dimensional. At any point, we can only find two mutually orthogonal lines, such as x_1 and y_1, or x_2 and y_2.

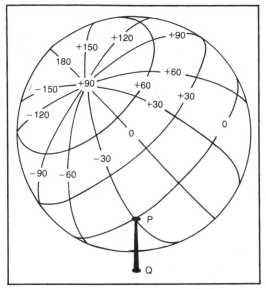

Fig. 6-3. The position of a point on the surface of a sphere can be determined by using the familiar latitude and longitude system. Longitude angles range from −180 to +180 degrees, and latitude angles range from −90 to +90 degrees. Point P is at latitude zero, longitude −30. By adding a radial distance coordinate, emanating from the center of the sphere, we can define all points in space. Point Q, for example, has the same latitude and longitude numbers as point P, but Q is farther from the center of the sphere.

dinate system. This system gets its name from the French mathematician Rene Descartes, who first invented and used it. It is generated by constructing sets of mutually perpendicular number lines.

Figure 6-4 shows the evolution of a Cartesian system of coordinates for two and three dimensions. We begin with a simple real number line, as shown at A; by placing two such number lines together so that they intersect at a common point (zero for both lines), and at a right angle, we obtain the Cartesian coordinate plane for two dimensions (B). This plane is the one you see in most graphic illustrations, such as rate-of-inflation charts or quarterly profit reports. Any point on the Cartesian plane is defined by two numbers. In Fig. 6-4, the point P corresponds to x = 3 and y = −2, and this is represented by an ordered pair of numbers, (3,−2).

In Fig. 6-4C, we show the three-dimensional Cartesian coordinate system. Here, the three

number lines, called the x, y and z axes, are mutually perpendicular at the point x = 0, y = 0, z = 0, defined by the ordered triple (0,0,0). The angle between the x and y axes is exactly 90 degrees; the same holds true for the y and z axes, as well as the x and z axes. Any point in three-space can thus be represented by a unique ordered triple.

We can project the Cartesian system into any number of dimensions. For n-dimensional hyperspace, the Cartesian system requires exactly n mutually orthogonal lines, and exactly n coordinate values. We call these lines the $x_1, x_2 \ldots x_n$ axes, and any point is represented by an ordered n-tuple ($x_1, x_2 \ldots x_n$).

POLAR COORDINATES

There are other ways to establish the uniqueness of a point, of course, besides the Cartesian system. On the spherical surface of the earth, for example, a Cartesian system is not very convenient because of geometric distortion. While it will work just fine for a small area, things get rather messy if we try to use the Cartesian scheme over the whole surface of a sphere.

To locate geographic points on the surface of the earth, then, we have adopted a system using angles instead of number lines. These angles are, of course, the familiar latitude and longitude coordinates. Figure 6-3 illustrates this kind of system. To define points in three-space, we add a radius coordinate, as shown. Then, any point in space can be uniquely represented by two angles and a distance vector.

Figure 6-5 illustrates a second kind of three-dimensional polar coordinate arrangement, using one angle and two distance values. To locate a given point P, we need to know the angle θ with respect to a given reference line, and a radius r, which determines a point P′ on the "image plane" that lies directly above or below P. Then we find the distance d from the "image plane" to P.

From the polar or Cartesian three-dimensional systems, we can generate a four-dimensional coordinate system by adding another distance vector. From the four-dimensional system, we can expand to five and more dimensions by adding progres-

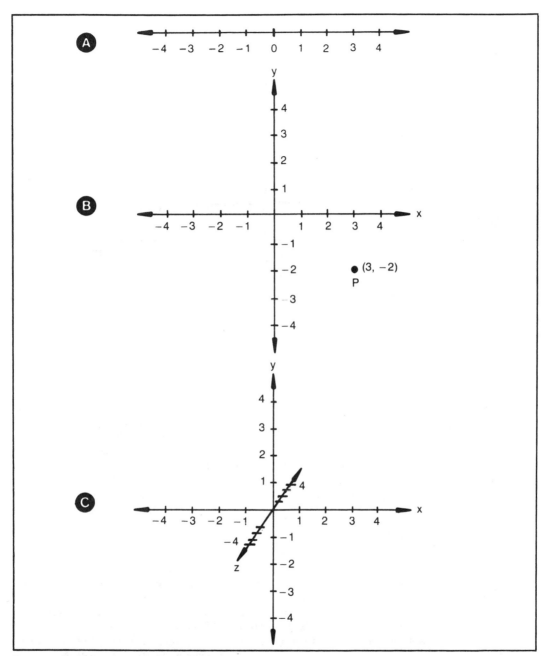

Fig. 6-4. At A, the basic number line is shown. By combining two of these at right angles, we obtain the two-dimensional Cartesian plane (B). By combining three such number lines at right angles, we obtain the three-dimensional Cartesian system (C). We can continue this process into any number of dimensions.

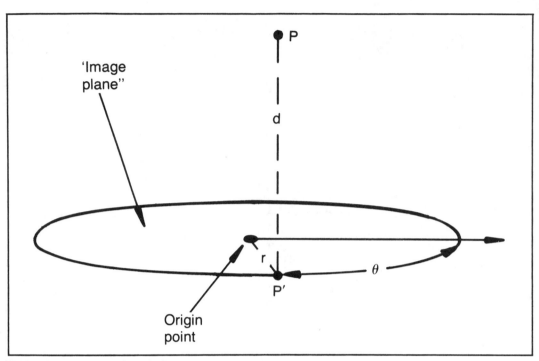

Fig. 6-5. A second polar method of defining points in three-space. An angle θ is specified with respect to the arrowed line; then a radius r and height d are given.

sively more distance vectors. The process can go on without end.

Visualizing a four-dimensional system of coordinates is, as we have said, essentially impossible for us three-dimensional beings. However, we can get some idea of the properties of hyperspace by considering time as the fourth dimension. We'll have more to say about this later.

TWO-SPACE

Henceforth, we will use the Cartesian coordinate system, rather than either of the polar versions, since the Cartesian scheme is geometrically easier to deal with. While the polar systems are equally valid from a mathematical standpoint, and are more convenient in certain applications, it is simpler to define distances and volumes with the Cartesian system.

Imagine what an existence would be like in a two-dimensional universe. For one thing, ge-

ometry would be considerably simpler than we know it; there would be no volumes, no "cubic" quantities, to be concerned with. We would have to worry only about length and area. But such a life would be restrictive compared to our familiar three-dimensional existence! A circle or square would completely imprison us (Fig. 6-6).

Depth perception would be possible in two-space, just as in three-space; binocular vision would allow two angles of viewing for any object. Through either eye, we would see an object as a straight line. If the lines, seen through either eye, happened to be identical in length, we would have no depth perception; but if they were different, we might imagine distance (Fig. 6-7). The distance to a circular object would be impossible to determine. Irregularly shaped objects, however, might be seen as being a definite distance away. It would greatly help to have a constant fog!

In two-space, the Cartesian coordinate system

92

may be used to determine the position of a point with respect to a selected origin. The two-space may or may not be "flat" as we three-dimensional beings envision it, but creatures confined to the continuum would perceive it as "flat." (They would not have the ability to perceive a third dimension, just as we cannot perceive a fourth.) We may define area as well as length in two-space. Objects will take up a certain number of square inches or feet or whatever units we happen to use. We determine areas in two space just the same way as we do in familiar plane geometry. To find the area of a square, we simply determine the length of a side and multiply it by itself, obtaining a certain number of square units. To find the area of a circle—the set of points equidistant from a common center—we find the radius r, and then multiply r^2 by π (3.14).

All areas, in fact, are determined in the normal way.

Of course, it is possible that a two-space continuum might not be "flat" in three dimensions. The surface of a sphere is a good example of a "curved" two-space. However, if we were confined to a spherical two-space, as two-dimensional beings, we would envision our universe as "flat." The whole idea that the universe might be closed—finite but unbounded—might very well never even occur to us, especially if the sphere were very large. We might never imagine that there could possibly be more than two dimensions, unless we were very astute mathematicians.

THE DISTANCE FORMULA FOR PLANE GEOMETRY

In two dimensions, the distance between two points on the Cartesian plane is quite simple to determine.

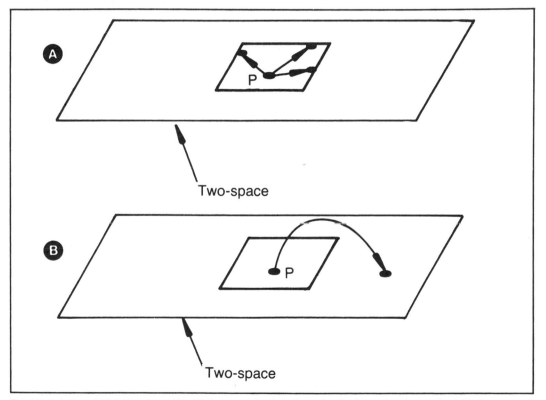

Fig. 6-6. At A, the point P is "imprisoned" inside the square in a two-dimensional universe. In order to get outside the square, P must travel through three dimensions, as shown at B.

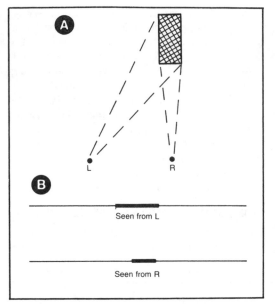

Fig. 6-7. An object, seen from two different angles in a two-dimensional space, may appear to have variable size. Seen from the left eye of a hypothetical two-dimensional being, the rectangle seems longer than it does through the right eye. At A, we show the situation as viewed from above the two-space; the left eye is labeled "L" and the right eye is labeled "R." At B, we show the respective views of the object, as seen by either eye. This effect would provide a sort of depth perception for a being in two-space, but only for certain objects seen from certain angles.

Suppose we have two points (x_1, y_1) and (x_2, y_2), as shown in Fig. 6-8A. Finding the distance between them involves the use of the theorem of Pythagoras for a right triangle, as shown. One side of the triangle has a length of $x_2 - x_1$, and the second side has a length of $y_2 - y_1$. The distance between the two points is the length of the longest side of this right triangle. The length of the hypotenuse, or longest side, of the triangle, is given by d, and:

$$d^2 = (x_2 - x_1)^2 + (y_2 - y_1)^2$$

Therefore:

$$d = \sqrt{(x_2 - x_1)^2 + (y_2 - y_1)^2}$$

Of course, d must be measured in whatever units we use to establish the coordinates of the two points. (We cannot use inches as our coordinate units, and then expect to find the distance d in centimeters.)

A simplification of this distance formula is used to find the diagonal of a unit square, or a square that

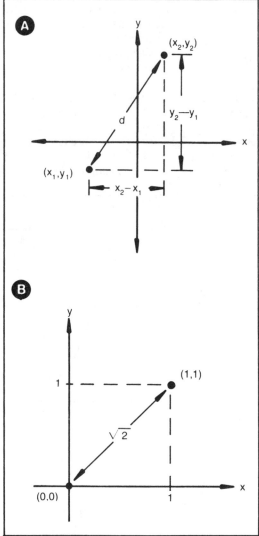

Fig. 6-8. The distance between two points on the coordinate plane is found by using the Pythagorean theorem for a right triangle. Here, $d^2 = (x_2 - x_1)^2 + (y_2 - y_1)^2$, for two arbitrary points (x_1, y_1) and (x_2, y_2), as shown at A. At B, the diagonal of a square is found by using the distance formula in simplified form.

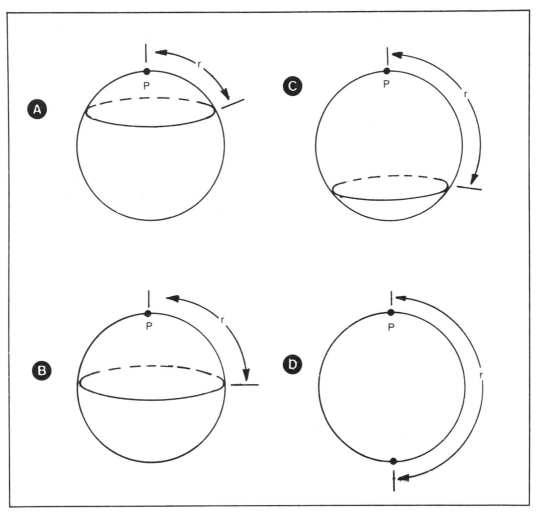

Fig. 6-9. When a two-dimensional continuum is non-Euclidean, such as the surface of a sphere, the perimeter of a circle is no longer a simple function of its radius. At A, we see that the circle must have a perimeter smaller than $2\pi r$. At B, the circle is as large as it can possibly be in the spherical two-space, since its radius is one-quarter of the circumference of the entire universe. At C, we see that further increasing the radius will cause the perimeter of the circle to become smaller, not larger. If the radius is made to be half the circumference of the spherical universe (D), the perimeter of the circle will vanish.

measures one unit on each side. In this particular case, we can consider that $x_1 = 0$, $y_1 = 0$, $x_2 = 1$ and $y_2 = 1$, as shown at Fig. 6-8B. Then the diagonal distance, from (x_1, y_1) to (x_2, y_2), is $\sqrt{2}$ units.

EUCLIDEAN OR NON-EUCLIDEAN?

In two dimensions, the "surface" of an object is just its perimeter, if we define "surface" to mean the boundary between the interior and exterior of an object. The perimeter of a square having a side of length s is just 4s; the perimeter of a rectangle with sides of lengths s_1 and s_2 is simply $2s_1 + 2s_2$; the perimeter of a circle of radius r is $2\pi r$, and so on. Of course, all these simple formulas are valid for a

95

"flat" two-space, but on the surface of a "curved" two-dimensional continuum, they will not be accurate.

Consider the example of a spherical two-space universe. When a circle on the surface of this sphere is very small by comparison with the sphere itself, we may use the formula $p = 2\pi r$ to obtain the perimeter of the circle, because the universe is essentially "flat" within a very small area. But as the radius r gets larger, the perimeter p will increase more slowly than it would in a plane (Fig. 6-9). When it becomes exactly one quarter of the circumference of the whole universe sphere, the circle will be a great circle with respect to that sphere, and this is as large as any circle can possibly be in the spherical continuum. Increasing the radius past this value, the circle will get smaller, not bigger; ultimately, when r gets to be half the circumference of the sphere, the circle will vanish to a geometric point. The radius could be increased yet more, and the circle would begin to grow again. This same sort of thing would happen to any object in a spherical two-space.

If the simple, plane-geometry perimeter formulas always hold, precisely, in a given continuum, then we say that the continuum is Euclidean. This term comes from the name of the ancient mathematician, Euclid, who invented the elementary laws of plane geometry. A Euclidean two-space is, of necessity, a geometric plane. It would appear "flat" as seen from a three-dimensional point of view. Any two-space that deviates from perfect "flatness" is called non-Euclidean, since the laws of plane geometry no longer apply.

How would two-dimensional creatures, confined to their two-space and not even able to imagine what a three-dimensional existence might be like, be able to tell whether or not their universe was Euclidean? They would see a square as a square, and a circle as a circle. There would not be any apparent visual distortion. But if a circle with radius r did not always have a perimeter of exactly $2\pi r$, the two-dimensional creatures would know that something was a little strange. If a small square had a different ratio of perimeter to area than a large one, our little plane-universe beings would eventu-

ally discover the reason. We, in our three-dimensional space, have discovered that our universe is non-Euclidean. Usually, the deviation from the "flatness" of Euclidean geometry is not very great, but under certain circumstances, such as when gravity is very intense or distances very large, our universe is far from Euclidean, indeed!

EUCLIDEAN AND NON-EUCLIDEAN THREE-SPACE

The distance formula for three-dimensional Cartesian coordinates is an extension of the two-dimensional case. For simplicity, let us imagine that we have two points, (x_1, y_1, z_1) and $(0,0,0)$. We must first find the diagonal of the "base rectangle" of the rectangular prism shown in Fig. 6-10. This "base rectangle" lies in the xy-plane, and its diagonal is the distance from $(0,0)$ to (x_1, y_1) in this plane. We already know the distance formula for two dimensions; therefore the diagonal d' of the "base rectangle" in the xy-plane is:

$$d' = \sqrt{x_1^2 + y_1^2}$$

To find the distance from $(0,0,0)$ to (x_1, y_1, z_1), we need to find the length of the hypotenuse of the right triangle shown, with vertices at $(0,0,0)$, $(x_1, y_1, 0)$ and (x_1, y_1, z_1). We already know the lengths of the lesser sides: One is z_1 units (the height) and the other is $\sqrt{x_1^2 + y_1^2}$ units (the base). We can use the Pythagorean theorem to find the desired distance, d:

$$d = \sqrt{(\sqrt{x_1^2 + y_1^2})^2 + z_1^2}$$
$$= \sqrt{x_1^2 + y_1^2 + z_1^2}$$

According to this formula, then, the diagonal of a unit cube, measuring one unit on each edge, is $\sqrt{3}$ units. The volume of a cube is obtained, of course, by finding the length of an edge and raising this value to the third power; a cube with edges of length s thus has a volume of s^3. The surface area of such a cube is $6s^2$, since a cube has six identical square faces.

You should recall from geometry courses that the surface area of a sphere having radius r is $4\pi r^2$, and the volume is $4\pi r^3/3$. These formulas are, of course, all based on the assumption that space is Euclidean. A square *looks* square, and a cube *looks* like a cube. There is no apparent distortion. The idea that our three-space could be anything but Euclidean seems somewhat analogous to the thought that we might divide by zero! What could possibly be meant by "non-Euclidean space"?

Imagine that we were to conduct an experiment using cubes and spheres of various sizes, and we checked the accuracy of the surface-area and volume formulas we have just identified. Suppose we had access to instruments of extreme accuracy, capable of measuring distances equivalent to billionths or trillionths of the diameter of a proton.

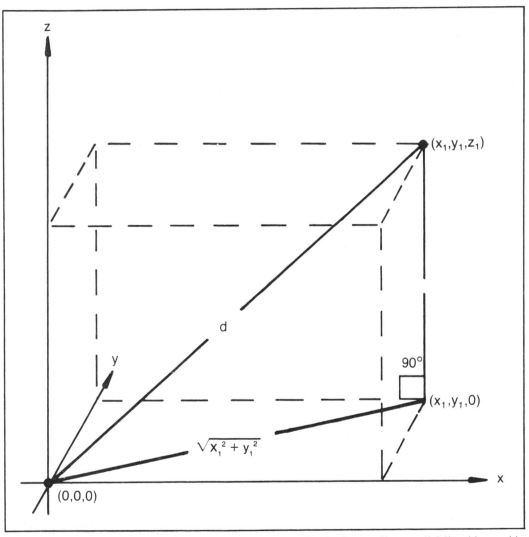

Fig. 6-10. Derivation of the distance formula for three dimensions. We find the distance d between (0,0,0) and (x_1,y_1,z_1) by constructing a right triangle whose lesser sides are already known. Then, we can apply the theorem of Pythagoras to obtain the value of d.

Suppose that our spheres and cubes could somehow be made perfect. We would find that our formulas are in fact not quite perfect.

How could they not be perfect? After all, they are derived mathematically! The answer must be, of course, that three-dimensional space is not Euclidean.

This borders on the unimaginable. Physicists of just a century or two ago would have laughed at the idea that our universe is actually this way. Galileo and Copernicus received similar reactions from the Establishment. It is gravity that causes space to be non-Euclidean, and we will discuss this effect in more detail in Chapter 7.

It is thought that our universe might in fact be a "hypersphere" or "four-sphere"—a four-dimensional set of points, equidistant from some cosmic epicenter—of huge proportions. Local deviations from the perfection of Euclidean geometry are very, very small; but many cosmologists think that, if we were to travel far enough in one direction, we would eventually arrive back at earth from the opposite direction. The circumferential distance around the "hypersphere" universe is probably on the order of tens or hundreds of billions of light years, so such a trip is not too practical. (Even at relativistic speeds, we would arrive back in the solar system to find the sun dead and our home planet either vaporized or frozen.)

THE HYPERSPHERE

Let us briefly examine some of the properties of a universe that is shaped like a "hypersphere." First, we ought to define the term "sphere" as it applies to a continuum having any number of dimensions. We specify an epicenter point, P; then an "n-sphere" is the set of all points in n-space at a fixed distance r from P. In a one-dimensional space, then, a one-sphere consists of only two points. A two-sphere is a circle, and a three-sphere is what we generally envision when we think of the word "sphere." A one-sphere has a surface with zero dimensions, since points are dimensionless. A two-sphere has a one-dimensional surface, and a three-sphere has a two-dimensional surface. In general, an n-sphere

has a surface of n − 1 dimensions. Our universe, if it is shaped like a four-sphere, has a three-dimensional surface, on which we exist. We never see anything inside or outside the four-sphere. We only see what is on the surface, since light travels only over the surface.

If we were to construct three-spheres of ever-increasing radius on the surface of a four-sphere, we would find that the three-spheres would keep getting larger and larger, in terms of volume and surface area, for a while. But eventually they would begin to reach a limiting size, and then start getting smaller. As the spheres grew smaller, they would seem to converge on a new center, located exactly on the other side of the four-sphere universe. Figure 6-9 is a dimensionally reduced illustration of this situation. When the largest possible three-sphere is constructed on the surface of a hypersphere, the three-sphere may be thought of as having two centers. In fact, any three-sphere really has two centers with respect to a hyperspherical universe: a "near center" and a "far center", located exactly opposite each other on the four-sphere.

While all this is easy to see in Fig. 6-9, it is rather hard to imagine how it could actually take place with another dimension added! But scientists believe our universe might be like this. As such, the universe would have no limits, no points beyond which we could not pass—but there would be only a finite number of cubic light years to explore.

THE FOUR-CUBE

Consider a four-dimensional system of Cartesian coordinates. Such a system has four number lines, all intersecting at a common origin point where they are mutually perpendicular. We cannot draw this coordinate scheme, but we can describe it in mathematical terms. We might call these number lines the w, x, y and z axes, and let them intersect where all four variables w, x, y and z are equal to zero. Any point is thus defined by an ordered quadruple (w, x, y, z).

Suppose we choose all the points in this coordinate system where each number is equal to either zero or one. There are 16 such points, as follows:

$$P_1 = (0,0,0,0) \qquad P_9 = (1,0,0,0)$$
$$P_2 = (0,0,0,1) \qquad P_{10} = (1,0,0,1)$$
$$P_3 = (0,0,1,0) \qquad P_{11} = (1,0,1,0)$$
$$P_4 = (0,0,1,1) \qquad P_{12} = (1,0,1,1)$$
$$P_5 = (0,1,0,0) \qquad P_{13} = (1,1,0,0)$$
$$P_6 = (0,1,0,1) \qquad P_{14} = (1,1,0,1)$$
$$P_7 = (0,1,1,0) \qquad P_{15} = (1,1,1,0)$$
$$P_8 = (0,1,1,1) \qquad P_{16} = (1,1,1,1)$$

These 16 points represent the vertices of a "unit four-cube," or a unit cube in four dimensions. We cannot visualize this object directly, because of the fact that we are not four-dimensional beings! But we can geometrically discover some of its properties.

Attempts to crudely illustrate a four-cube have led to drawings similar to Fig. 6-11. Here, a four-cube is shown as a cube within a cube. Actually, a true four-cube has edges that are all the same length, so this three-dimensional representation (further reduced to a two-dimensional perspective drawing to make it fit on a flat page) is distorted. A four-cube has 16 vertices, while a three-cube has

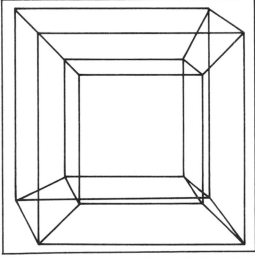

Fig. 6-11. The four-cube may be illustrated in distorted form as a three-dimensional model, here reduced to a two-dimensional perspective drawing. All of the straight-line edges are actually the same length in a real four-cube. Here, the model appears as a cube within a cube.

eight; a four-cube has 32 edges, while a three-cube as 12.

The diagonal length of a unit four-cube is obtained from the distance formula for four-dimensional Cartesian coordinates. It is actually quite easy to derive this formula, but the pictorial illustration is horribly difficult to construct. A little imagination will go a long way here!

Suppose we wish to find the distance between the two points $(0,0,0,0)$ and (w_1,x_1,y_1,z_1) in the wxyz-hyperspace. We can use the formula for the three-dimensional distance through one of the cubical "faces" of the hypercube, and then use the theorem of Pythagoras, in a manner exactly like we did to get the diagonal distance through the three-cube in Fig. 6-10. There are four three-dimensional subspaces within the wxyz-hyperspace; they can be defined by setting each of the coordinates, in turn, to zero. This gives us a choice among the spaces $(0,x,y,z)$, $(w,0,y,z)$, $(w,x,0,z)$ and $(w,x,y,0)$. Let's use the $(0,x,y,z)$ subspace.

The distance between $(0,0,0,0)$ and $(0,x_1, y_1,z_1)$ is already known to us; we use the three-dimensional distance formula to obtain this distance as:

$$d' = \sqrt{x_1^2 + y_1^2 + z_1^2}$$

We may construct a right triangle, using the technique of Fig. 6-10 applied to one extra dimension. The two lesser sides of this triangle will have lengths d' and w_1. Then, by the Pythagorean theorem:

$$d = \sqrt{(\sqrt{x_1^2 + y_1^2 + z_1^2})^2 + w_1^2}$$
$$= \sqrt{w_1^2 + x_1^2 + y_1^2 + z_1^2}$$

In the special case of a unit four-cube, where $(w_1,x_1,y_1,z_1) = (1,1,1,1)$, the length of the diagonal is $\sqrt{4}$, or 2, units. You may recall that a square has a diagonal length of $\sqrt{2}$ units if its sides are one unit long, and a similar cube has a diagonal of $\sqrt{3}$ units. Apparently a unit n-cube will have a diagonal length of $\sqrt{n}$, and we can prove this without much difficulty.

The four-volume, or hypervolume, displaced by a four-cube is equal to the fourth power of the length of one edge. In n-space, the n-volume of an n-cube is equal to s^n, where s is the length of one edge. For a rectangular four-prism, the hypervolume is given by wxyz, where each variable represents the length of the object in each of the four dimensions.

TIME AS THE FOURTH DIMENSION

In Chapter 3, we briefly looked at some coordinate-system models using time as the fourth dimension. From a theoretical standpoint, time is perfectly applicable in this role. We may think of time as constantly "flowing" or "moving" along a line, and we may assign coordinates to the line. This is often done in history textbooks to give us a better idea of the comparative times of, for example, the Civil War as opposed to the decline of the Roman Empire or the evolution of the solar system.

We may choose intervals of any magnitude for the coordinates of a time line. For the above example, we might use years or centuries. For other purposes we might want to choose days, hours, or perhaps minutes. For still other purposes we might use microseconds or nanoseconds (units of 10^{-6} or 10^{-9} second) as our base unit. Time is truly a continuum, in the sense that we can specify units on a time line in whatever magnitude suits us.

As we saw in Chapter 3, there are theoretical reasons why we should consider a correlation between time and spatial distance, based on the constancy of the speed of light. An interval of one second is equivalent, in a spatial sense, to a distance of 300,000 kilometers. An interval of one year, by the same reasoning, is equivalent to a distance of one light year, and so on. Table 6-1 shows the equivalents of several familiar units of time and distance according to this analogy.

According to this model, our three-dimensional space continuum may be imagined as traveling, with speed c, through the fourth dimension; the "time" or t axis is thus perpendicular to all the "space" axes in any Cartesian coordinate model. Figure 6-12 is a dimensionally reduced illustration of this; our three-space is shown as a two-space so we can comfortably fit the diagram into an illustration. Points P and Q are 300,000 km apart as mea-

Table 6-1.

Distance	Time
Diameter of baseball	0.23 nanoseconds
Height of man	5.9 nanoseconds
Length of football field	0.3 microseconds
Distance across USA	21 milliseconds
Diameter of Earth	42 milliseconds
Distance, Earth to Moon	1.3 seconds
Distance, Earth to Sun	8 minutes
Diameter of Solar System	10 hours
Diameter of Milky Way	100,000 years
Radius of known universe	10 billion years

sured through the space continuum. The whole three-space moves forward (or toward the future) along the t axis at a velocity such that it traverses a distance the same as PQ in exactly one second.

In the four-space thus created, all objects that we regard as three-dimensional become four-dimensional. In the dimensionally reduced drawing of Fig. 6-13, the circle becomes a cylinder without ends (A), and the square becomes a square prism without ends (B). In space of three dimensions, a circle and a square describe the same space-time objects as they would in two-space; but a sphere and

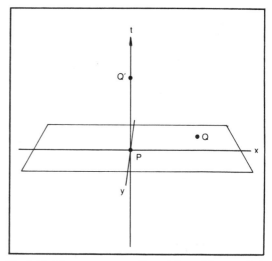

Fig. 6-12. A two-space moves through time at the speed of light, creating a three-space when we consider time as a dimension. Here, the distance PQ is the same as the distance PQ'. If PQ is 300,000 kilometers, for example, then PQ' is one second.

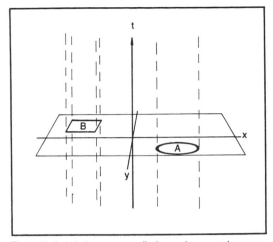

Fig. 6-13. A circle becomes a cylinder, and a square becomes a square prism, both without ends, when we allow time to be a dimension.

cube describe space-time objects that we cannot illustrate or imagine except in mathematical terms.

Imagine that, somehow, we could create and destroy certain two- and three-dimensional geometric objects at will. Suppose we call into existence a square whose edges each measure 300 meters in length. Further suppose that, exactly one microsecond after we have created this square, we destroy it. Since 300 meters of spatial distance corresponds to one microsecond of time, the object resulting from this creation-and-destruction event is, as Fig. 6-14 shows, a cube. It is a three-cube, but not the kind of three-cube we ordinarily imagine; instead of occupying three spatial dimensions, this three-cube exists in two spatial dimensions and one time dimension. We may call these x, y, and t. Or, if you prefer, we might call them x, z, and t, or perhaps y, z, and t. It makes no difference.

By a similar creation-and-destruction process, we can mentally form a three-sphere (Fig. 6-15). The sphere requires a bit more effort to mold; it is generated by first creating a point (A), then changing it into a circle (B), and finally reducing the circle again to a point (C) before extinguishing the figure altogether. But this must be very carefully done. If the circle does not change size at just the right rate, the resulting space-time object will not be a sphere,

but a badly distorted ellipsoid or blob. The circle must achieve just the right maximum diameter, corresponding to an equivalent period of time from the first appearance of the point (A) until its final demise (C). However, assuming all the parameters are correct, we will be able to generate a true three-sphere in space-time, although not the kind of sphere we ordinarily think of!

By generating an expanding and contracting three-sphere, with a maximum diameter corresponding to the period of its existence and having just the right rates of inflation and deflation, we can generate a four-sphere in space-time. By similar mental exercises, we can create ellipsoids, paraboloids, and other geometric objects in three and four dimensions. It is fortunate, however, that we are not faced with playing "God" in the way we have been describing! Some of the objects we have talked about are actually impossible to create in reality of true space-time.

In the cases of the three-sphere and four-sphere just described, you may have noticed that the initial and final speeds of expansion and contraction approach infinite values. At the very start and finish of the three-sphere or four-sphere generation, the circle or sphere must change size at a tremendous speed. But, the speed of light is the fastest possible velocity, and it is represented in the continuum we have built as a path having an

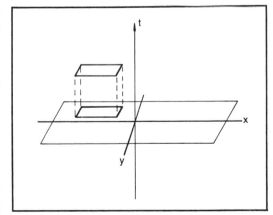

Fig. 6-14. By generating a square in two-space, having sides each 300 meters long, and then destroying it one microsecond later, we generate a cube in space-time.

inclination of 45 degrees to the t axis. In Fig. 6-15, we can see that the circle's points move with a slope greater than 45 degrees relative to the t axis. This cannot happen! Thus, such spheres as we have imagined must remain imaginary, not real. Space-time is a "limited" four-space in this sense. The t dimension is not amenable to free movement as are the other three.

ESCAPING FROM A CELL

Despite the shortcomings of labeling time as the fourth dimension, the space-time model of hyperspace does, at least, make the concept of a four-dimensional universe a bit easier to intuitively

swallow. Consider, for example, how easy it is for us to imagine how a two-dimensional creature would get out of a circular or rectangular enclosure. While the poor imprisoned two-space being feels totally confined by such a geometric set of walls, we might be equally frustrated at his ignorance of the third dimension, via which he could so easily get free if there were only some way for him to gain access to it! With respect to four-space, a three-dimensional prisoner in a sealed cubicle faces exactly the same predicament.

If travel through time were possible, it is easy to see how a prisoner might escape from a sealed cubical cell. All he would have to do is go back in

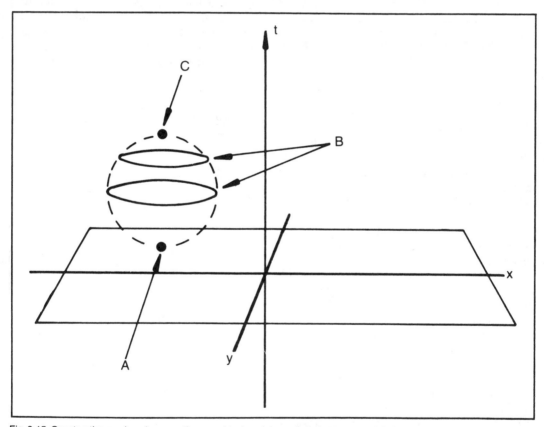

Fig. 6-15. Constructing a sphere in space-time requires imagining a circle that grows and shrinks at just the right rate and to just the right maximum size. A point appears initially, at A; this grows into a circle of changing size, at B. Finally, the circle shrinks back to a point, at C, and then disappears. If the rate of expansion and contraction of the circle is not just right, we will get an ellipsoid, or flattened sphere, or maybe just a formless blob.

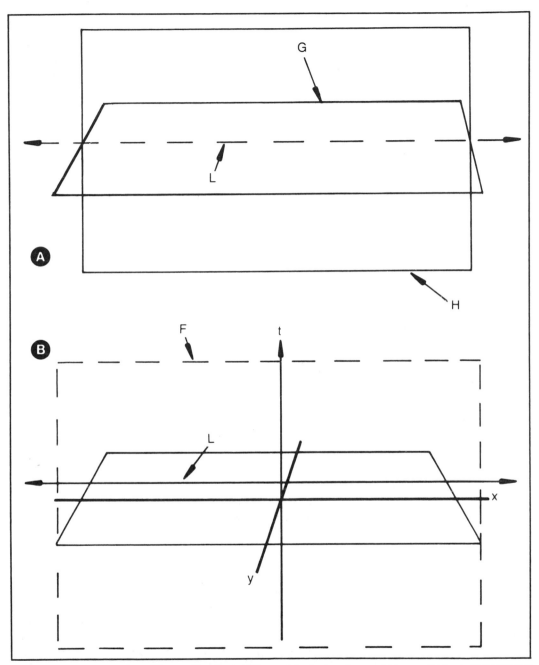

Fig. 6-16. At A, two planes G and H intersect in a line L. At B, supposing the two planes exist forever, the line L moves through time, forming a plane F, which is the intersection of two three-spaces created by planes G and H as they move through time.

time until before the building was constructed, move over a few feet, and then return to the present, which would find him outside the cell. Or, alternatively, he could travel into the future until the building crumbled with age or was torn down, move over a few feet, and travel back to the present. Such a time-travel excursion is geometrically equivalent to four-dimensional travel, wherein the cell would be quite easy to escape from.

Of course, time travel is not possible, at least not backwards, so this illustration is not very realistic. But, it does help us imagine how a creature capable of moving through four dimensions could escape from an apparently secure room. To us, it would seem as though the creature would disappear from the interior of the cell and then reappear outside. Whether-he had traveled through time, or through a spatial fourth dimension, would not be ascertainable, except by the logic that backward time travel cannot take place.

INTERSECTING SPACES

In three-space, we can easily visualize how two planes may intersect to form a line (Fig. 6-16A). In four-space, two three-spaces can intersect to form a plane. The space-time model of hyperspace allows us to illustrate this in the following way.

Imagine two planes, as in Fig. 6-16A, existing indefinitely; that is, as far into the past and future as possible. Then these two planes are, in the four-dimensional space-time continuum, three-spaces. If they intersect in a line as shown, then this line, of course, moves through time along with the planes, and so it describes a two-space, or plane, in the space-time universe. This is shown in Fig. 6-16B.

It is possible that the two planes might not intersect at all, in which case they are parallel (Fig. 6-17). Then the two planes traverse non-intersecting three-spaces through the universe of space-time. These are "parallel" three-spaces in a four-dimensional hyperspace.

Three-spaces generated in the space-time continuum do not, of course, have to be infinitely large, as lines or planes or unlimited spaces. Consider the case of an expanding and contracting three-sphere, represented in dimensionally reduced form by Fig. 6-15. Considering only the two-dimensional surface of this evolving "balloon," a non-Euclidean three-space is generated in the universe of space-time. This three-space is the surface of a four-sphere and is unbounded but of finite proportions.

Imagine two of these balloon spheres, expanding and contracting side by side. Suppose that they both come into existence at the same instant, and are close enough together so that they overlap when they expand to their maximum extent. This intersection (Fig. 6-18) will first be a point (A), but it will grow into a circle (B) as the spheres continue to expand. When the balloons reach their largest sizes, the circle of intersection will be at its largest. Then, as the spheres contract, the circle will get smaller and smaller, finally becoming just a point again (C), and then disappearing altogether. As we have shown them in Fig. 6-18, the two balloons appear and disappear simultaneously, and grow to exactly the same maximum size.

The intersection of these two three-spheres is thus either a point or a circle, and this expanding and contracting circle traverses space-time as shown in Fig. 6-15. The result is a three-sphere, of two spatial dimensions and the time dimension. This shows how two four-spheres can intersect to form a three-sphere.

The intersection of two four-spheres might be only a point, if the two balloons got only large enough to touch for an instant before contracting; or, if the balloons did not get large enough to touch at all, the four-spheres would be disjoint.

HOW WE SEE LESSER SPACES

We have described geometric lines, circles, squares, planes and cubes, as well as other objects, quite liberally. But what do perfect geometric lines or planes look like to us?

The answer is that such universes would be totally undetectable to us. A geometric line has length, but no width or depth. A geometric plane has, say, length and depth, but no width. Any object in a one-space or two-space would appear to us to have zero volume and zero mass. So there might be a two-space slanting right through you at this very moment, and you have no way to tell that it is there. You could not see it; all three-space photons would go through it without being affected, and two-space photons would have no energy by our reckoning. You would not hear it; our sound waves would pass right through, and their sound waves would remain confined to their universe, being vibrations of massless and infinitely thin molecules. Mathematically, a one-space or two-space can exist, but in the sense of our own reality, they might as well not. Whether or not these universes are there is of no consequence to our universe.

To any one- or two-dimensional being, we would appear to be of infinite proportions, assuming

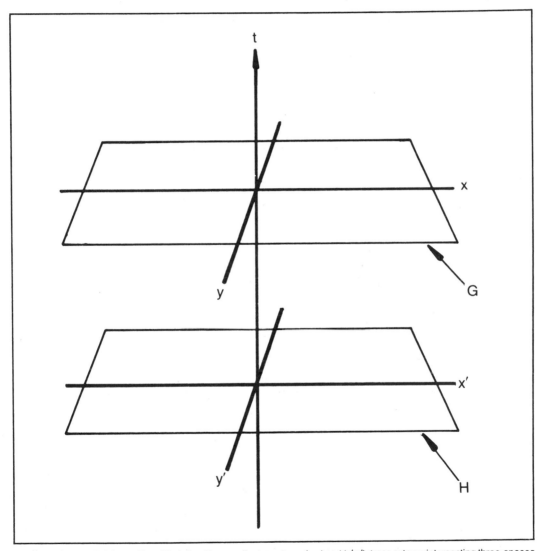

Fig. 6-17. Two parallel planes G and H, defined by coordinate systems (x,y) and (x′,y′), trace out non-intersecting three-spaces as time passes.

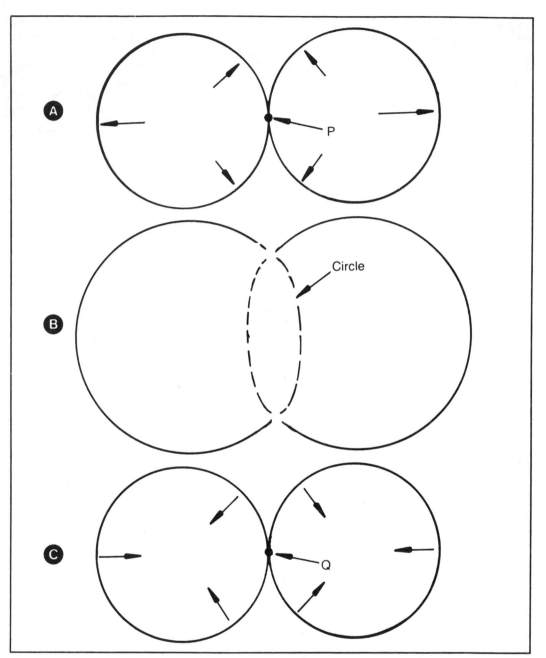

Fig. 6-18. Two identical three-spheres expand and contract. At A, their intersection is a point P; at B, this point grows to a circle; at C, the circle shrinks again to a point Q as the three-spheres contract. These three-spheres are four-spheres in the space-time continuum; the circle becomes a three-sphere, representing the intersection of the two four-spheres.

they were aware of our real existence. They would see us as cross sections in their limited realms, but they would be unable to distinguish (except perhaps by deductive reasoning) these cross sections from objects in their own universes. The drawing in Fig. 6-19 shows an example of this. Creatures in two-space see the pear as an object, but they have no way of knowing what the object truly is. If the object were to move, passing through the plane universe, it would appear first as a point, then as a roughly circular blob of varying size, and finally it would disappear again. The two-space beings would doubtless be shocked by this, but they might be smart enough to figure out what was actually happening.

Extending this concept to another dimension, suppose we were four-dimensional creatures. We would not be able to detect the existence of a three-dimensional subspace. But we are, in fact, four-dimensional when we let time be a dimension!

Looking at it this way, what is a truly three-dimensional object? It is either a two-dimensional spatial thing lasting a certain length of time, or else it is a three-dimensional spatial thing that lasts for no time at all. Either kind of object cannot be detected by us: the former is too "thin," and the latter is gone before we can tell that it was ever there.

Now, we can get some idea of what a creature of four spatial dimensions would see of us. The answer is nothing! But we might see them as strange objects that appear out of nowhere and then disappear again, like sorcerers. We should not be too surprised if we find celestial objects that behave this way. Some scientists believe that matter might come from another dimension and deposit itself in our universe, seeming to form like magic out of the intergalactic vacuum.

MORE THAN FOUR DIMENSIONS

Defining the fourth dimension as time is convenient

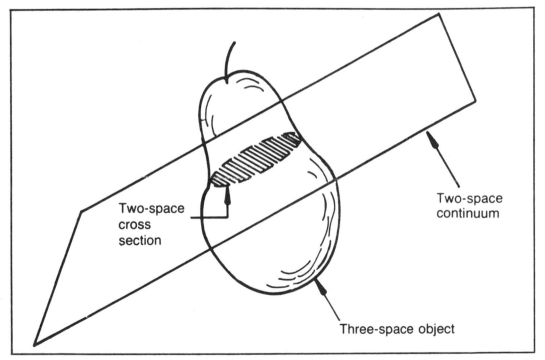

Two-space cross section

Two-space continuum

Three-space object

Fig. 6-19. A creature in a two-space continuum would see the cross section of the three-dimensional pear as a roughly circular object. He would not be able to tell whether the object was completely contained in his universe.

for our imaginations, but we have seen that there are certain shortcomings involved with this approach. The four-sphere, for example, cannot be defined without requiring speeds in excess of the speed of light. The best we would be able to do is give the four-sphere pointed poles. Therefore, we will henceforth not consider this definition of the "fourth dimension." From now on, when we speak of a dimension, we will mean to say a spatial dimension.

Geometrically, we can define cubes and spheres in any number of dimensions. An n-cube is an object occupying n dimensions and having 2^n vertices—points at which exactly n lines of equal length intersect in a mutually perpendicular way. In a coordinate system of n dimensions, we can generate the vertices of a unit n-cube by counting upwards in the binary number system until we reach 2^n. For example, in seven-space, there are 2^7 or 128 vertices in an n-cube. These points are represented by the ordered septuples (0, 0, 0, 0, 0, 0, 0), (0, 0, 0, 0, 0, 0, 1), (0, 0, 0, 0, 0, 1, 0), and so on up to (1, 1, 1, 1, 1, 1, 1). An n-sphere is simply the set of all points in n-space that lie a given radius r from a certain epicenter point P.

Is such a discussion meaningful? It may be mathematically intriguing, and can certainly get very complicated when we start imagining any but the simplest geometric forms. As mentioned before, cosmologists believe that our three-space continuum is non-Euclidean, or curved. For this to be possible, we must accept the existence of a fourth spatial dimension, so that our universe has something with respect to which it can be curved! In fact, there is no good reason why there should be any limit to the number of dimensions that are possible. A three-space can be imbedded in a four-space, which in turn lies within a five-space; in general, an n-space can be considered to be a subspace of an n+1-space. It is harder to imagine that there should be any limit to the possible number of spatial dimensions, than it is to think that there can exist an unlimited number of them.

The test for the existence of a further dimension, no matter how many dimensions a given universe may have, can be conducted by means of simple physical experiments. If we discover that a universe is non-Euclidean, then we can conclude that a further dimension must exist.

INTERIOR ANGLES OF A TRIANGLE

Suppose we construct a surveying apparatus, using lasers and instruments capable of precisely measuring distances and angles, as shown in Fig. 6-20. We might set up these instruments on the surfaces of planets, or on satellites. There must be three such stations, creating a huge triangle in space.

According to the laws of Euclidean geometry, we are taught that the sum of the measures of the three interior angles of any triangle is 180 degrees. But, this holds true only if the space under consideration, containing the triangle, is Euclidean. In a non-Euclidean space, the sum of the measures of the angles will not necessarily be 180 degrees. Figure 6-21 shows two possible instances where this law of Euclidean geometry is violated.

At A, we have a continuum with spherical or "positive" curvature, where the sum of the measures of the angles will be more than 180 degrees. At B, the continuum has a "negative" or saddle-shaped curvature, and this kind of non-Euclidean space will result in a total interior angle measure of less than 180 degrees. The drawings in Fig. 6-21 apply just as well to three-space as to two-space (as they are shown); in fact, this rule can be extended to apply to any n-space.

Suppose, then, that the triangle test yields results that show a non-Euclidean universe. It is possible to argue that the failure of the sum of the measures of the angles to be 180 degrees is not caused by curvature in the continuum. We might try to convince ourselves that space is really Euclidean, and the light beams from our surveying lasers are made to follow curved paths because of some external force, such as gravitation or refraction. In fact, both of these effects do cause light to deviate from a straight line. While we can explain away the effects of refraction, as we shall soon see, we cannot so simply dispense with gravitation.

By conducting experiments similar to the triangle test, we do find that space is non-Euclidean. There are no perfectly straight paths in our universe, but only "straightest" paths, consisting of the shortest distances between points. Light beams seem to us to travel in straight lines, but actually these paths are curved because the continuum itself is curved.

REFRACTION

When using light beams for checking whether or not space is Euclidean, we must consider refractive media. Suppose we set up the triangle-test apparatus in a swimming pool as shown by Fig. 6-22, so that two of the vertices of the triangle are under

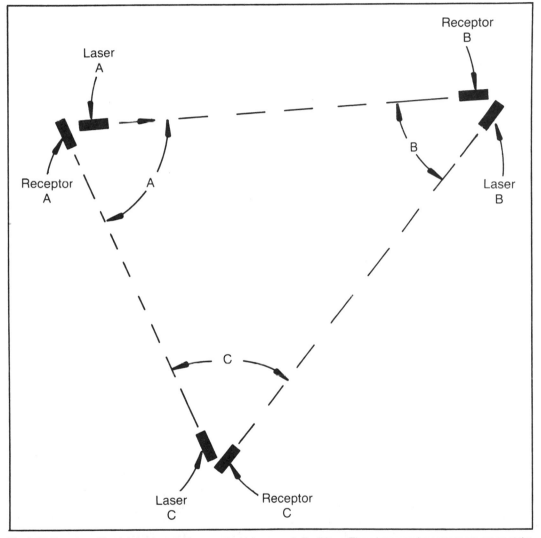

Fig. 6-20. Experiment for determining whether or not a given space is Euclidean. Three lasers and receptors are set up at the vertices of a triangle, as shown. If the space is Euclidean, the sum of angles A, B and C should be exactly 180 degrees. If the sum of the measures of the angles is not 180 degrees, then the universe is non-Euclidean. Several different orientations and sizes of the triangle might be necessary to ascertain if the space were "flat" or "curved."

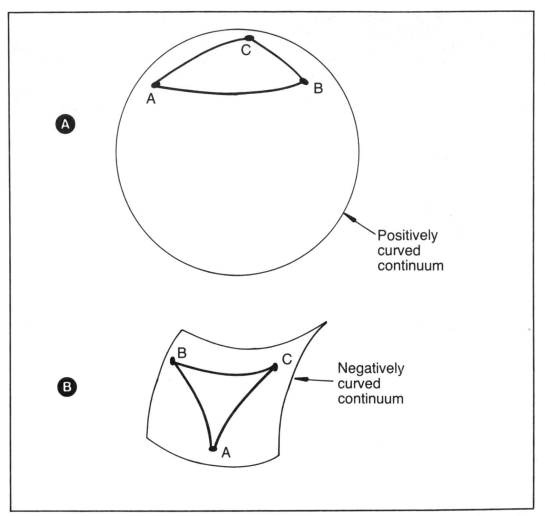

Fig. 6-21. Triangles on a positively curved surface (A) and a negatively curved surface (B). In the former case, the sum of the measures of the interior angles is more than 180 degrees; in the latter case, less.

water. When we measure the interior angles at the vertices of this "triangle," we will find that they add up to more than 180 degrees. This is because of refraction as the beam between A and C passes through the surface of the pool. The beam traveling between B and C is not bent since it hits the surface at a 90-degree angle, and of course the underwater path between A and B is not affected by changing refractive media.

But, if we drain the pool, we will find that the sum of the angles is different. Since the triangle is probably very small, and space is almost Euclidean within small regions, we would find that the sum of the measures of the angles was just about exactly 180 degrees. Does the addition of water to the pool make a nearly Euclidean space into a decidedly non-Euclidean one?

We might stretch a piece of metal wire be-

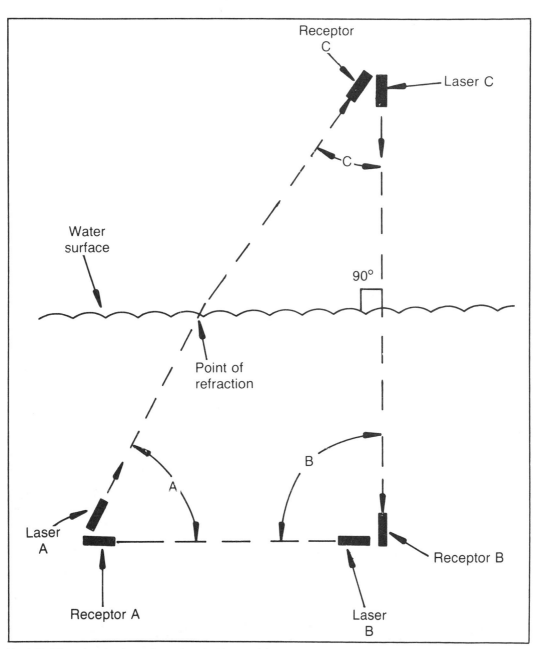

Fig. 6-22. When the triangle test is conducted with part of the apparatus submerged in a pool of water, we might get an indication that space is non-Euclidean (the sum of angles A, B and C will exceed 180 degrees). Draining the pool, we would find a difference in the measured sum of the angles. Surely this does not mean that water makes a difference in the curvature of the universe!

tween vertices A and C of the triangle in Fig. 6-22, and then we would find that the addition of water to the pool would have no effect on the straightness of side $\overline{AC}$. When the pool is full, the light does not travel the shortest route between A and C. We apparently cannot always rely on light beams to tell us what the shortest path between two points is. We have to be sure there is nothing in the way to cause refraction. Changing the frequency of the light beams, for example to extremely short gamma rays, would just about eliminate the refraction caused by the water. Surely the Euclidean or non Euclidean nature of space does not depend on the frequency of the lasers used for the triangle test!

In deep space, so far as we know, there are no changes in the index of refraction for light waves. If there were, there would be spectral blurring, since red waves would be bent to a different extent than blue ones. Therefore, we do not have to be concerned with refraction on this scale. But, is it possible that there is some other effect that bends light? There is; it is gravitation, and we shall investigate the effect of gravity on space more fully later.

It is much easier to explain astronomical phenomena in terms of non-Euclidean space, when we observe the deflection of electromagnetic beams and other objects, instead of attempting to postulate the existence of some unidentified refractive medium or force. Not only are light beams affected; we find that the non-Euclidean model of space, as predicted by Einstein in his general theory of relativity, explains quite nicely a quirk in the orbit of the planet Mercury, that was unexplainable before.

GEODETIC LINES

In any Euclidean or non-Euclidean universe, the line representing the shortest possible path between two points is known as a geodetic line. Of course, in a Euclidean universe, a geodetic line is "straight." In non-Euclidean space, there may occasionally be certain special cases where some geodetic lines are straight, but in general they will be curved. Consider our own earth. Imagine you are located in Chicago and have a radio receiver with a highly directional antenna system, and you want to point the antenna so that you will pick up signals from South Africa. You can stretch a piece of wire or string over the surface of a globe and find the great circle path to the desired location. This is the route an airliner would take, and it is also the route (approximately) that radio waves take. This line is obviously not straight. The radio waves cannot take a straight path from South Africa to Chicago. They would have to penetrate many miles of solid rock and metal!

On a spherical continuum, such as the surface of the earth, all geodetic lines are great circles—that is, they are parts of some circle whose geometric center lies at the center of the sphere. On a less symmetrical curved surface, geodetic lines are not as easy to define. Suppose two men stand near the summit of a hill, which has a rather irregular shape. How would they find the geodetic line connecting their respective positions? The simplest method would be for them to take a non-elastic rope, wire or chain, and pull it over the ground until it was as taut as possible. Assuming the rope did not get caught on an obstruction, and that the hill did not have any irregularities so severe that the rope might get oriented the wrong way, the rope, wire or chain would lie along a geodetic line according to the surface of the earth in that location. This is shown by Fig. 6-23A.

If the two men were to stand on different hills, for example at opposite sides of a valley (Fig. 6-23B), how would they find the geodetic line connecting their positions? They could use the same nonelastic rope, wire or chain, and lay it on the ground. Then they could pull it tight. At a certain point, further pulling would cause the device to lift up off the earth; just at this tension, the rope, wire or chain would lie along a geodetic line.

In the vacuum of intergalactic space in our universe, we have only beams of light to use as geodetic reference lines. Assuming no refraction, all beams of light move along geodetic paths through our universe. These lines are not necessarily straight; as a matter of fact, they essentially never are.

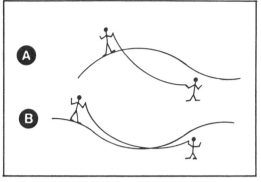

Fig. 6-23. Measuring for the location of a geodetic line on top of a hill (A) and in a valley (B).

COORDINATES IN "SPHERICAL" UNIVERSES

In non-Euclidean space, it is not possible to have a Cartesian coordinate system, because the axes have to be straight as well as mutually perpendicular at the origin. This must be true throughout the universe. Clearly, on the surface of a sphere, for example, this cannot be the case.

Consider the surface of a three-sphere universe, such as the surface of a planet. Imagine that the only method we have for determining the location of a point on this sphere is by means of geodetic lines. In a continuum where light can travel without obstruction, we might have access to lights and angle-measuring devices, and perhaps some way to determine distances. How can we establish the positions of points in the sphere-universe? One method is shown in Fig. 6-24A. We can choose two "poles" for the locations of light receptors. Any two places on the sphere will suffice, as long as they are not exactly opposite each other. By placing a light source on object P (or in case P happens to be a star it gives off its own light) we are able to see P from both of the angle-measuring stations X and Y. We first decide which direction will represent a zero-degree angle from either station, and then we can measure the angular position of P. We see that P has coordinates of, say, $(x,y) = (140,70)$, where the values are given in angular degrees.

We can move object P anywhere along the line $x = 140$, and the angle as measured from X will not

change. We can likewise move P anywhere along the line $y = 70$ without changing the angle as seen from Y. The lines $x = 140$ and $y = 70$, and in fact $x = a$ and $y = b$ for any constants a and b, are geodetic lines. We might call this coordinate scheme a system of "dual geodetic coordinates."

The one flaw in this system is that we may actually have two different points on the sphere that

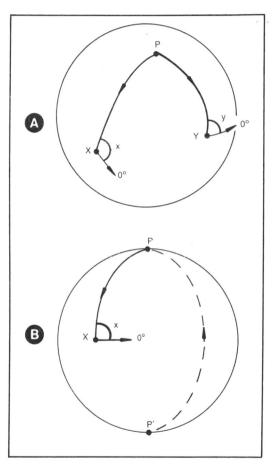

Fig. 6-24. The dual geodetic line method of generating a coordinate system on the surface of a sphere. We measure the angles x and y from points X and Y, determining the location of point P, shown at A. But there might be some confusion (B); light from P', exactly opposite P on the sphere, can travel the long way around the sphere and arrive at point X from the same angle as the light from P.

113

correspond to a given ordered pair of angles (x,y), as shown in Fig. 6-24B. Not only will point P appear in the position (x,y), but so will point P', located exactly opposite from P on the surface of the sphere. Also, the positions of both P and P' will correspond to not one, but two angles, from both stations X and Y. As well as lining up with the lines determined by the pair (x,y), they will also fall into line with a pair of angles (x',y'), both differing by 180 degrees from x and y. That is, we will see both points from opposite directions at the same time. We must have some way to establish distances if we are to get rid of this problem with a dual geodetic coordinate system.

A second method of defining the positions of points on a sphere is shown in Fig. 6-25. We first establish one pole, located at point X, and then we measure the angle x toward the object P, and the distance y as well. We must know how large the spherical universe is, if we are to avoid duplication of coordinates. Once we know how large the sphere is, we set the limitation that y must be no larger than half the circumference of the sphere. The resulting coordinate system is very similar to the latitude and longitude system we use on earth. Once we know the maximum permissible length for y, we can convert it to an angle having a range of zero to 180 degrees by means of a linear function. This coordinate system is also a geodetic system, but it involves only one geodetic line instead of two, and only one measuring station. We might then call this a "single geodetic system" of coordinates. Unfortunately, measuring huge distances from one point is not easy. We could perhaps bounce light rays off P if it were within a reasonable radius of our station X, but if P were millions of light years away, we would have to take an educated guess as to its distance from us.

Consider now the situation we face in our three-dimensional space, when we wish to set up a system of coordinates on the surface of a four-sphere. We cannot obtain two measuring stations separated by a very great distance, so the dual geodetic system is not practical. A set of geodetic coordinates of this sort would require three stations instead of two, anyway, in three-space; this would

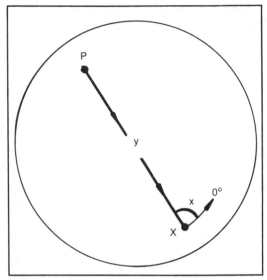

Fig. 6-25. The single geodetic line method of generating a set of coordinates on the surface of a sphere. We choose one point X, and measure the angle x and distance y to a given point P. To avoid duplicity, we set the maximum limit for y to be one-half the circumference of the spherical universe.

compound the difficulty. We must use a system that requires only one measuring station: earth.

We might use two angles and one distance vector, such as the celestial latitude and longitude and the estimated distance in light years. A better system, independent of the rotation of the planet, would make use of the celestial right-ascension and declination angles. We do not know exactly how large our "universe-sphere" might be, if it is really shaped like a four-sphere, but we must assume it is big enough so that we do not see one object as two objects from opposite directions.

COORDINATES IN "NON-SPHERICAL" UNIVERSES

The n-sphere lends itself conveniently to establishing a system of geodetic coordinates. But suppose the universe is not spherically shaped. Suppose that the surface of the universe is negatively curved (saddle shaped). Geodetic lines may still be used to determine coordinates, and we may still use a system of angles and distances. But, we may also consider a grid type setup. It is belived by cos-

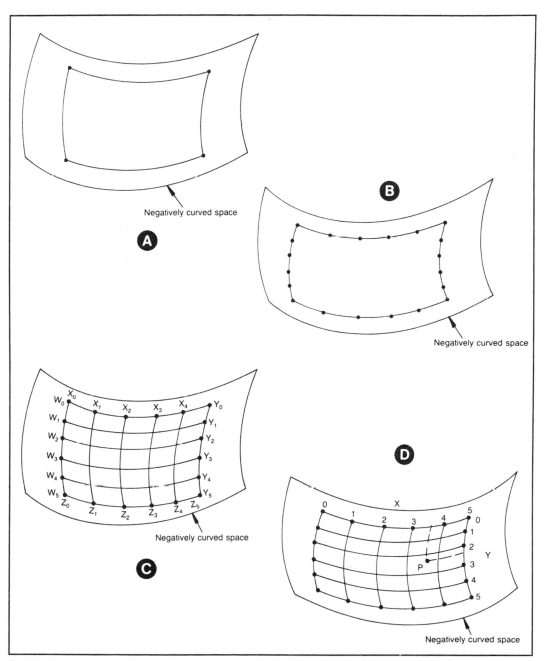

Fig. 6-26. A system of coordinates on a negatively curved continuum. We begin with four points (A), connecting them with geodetic lines. Then we partition these lines into equal segments (B). Next, we assign coordinate values to the points (C). The location of a point P may then be found (D).

mologists that our own universe is negatively curved in small regions near gravitational masses, but is positively curved on an interstellar or inter-galactic scale. (We will have more to say about this in Chapters 7 and 8.)

A localized set of non-Euclidean two-space coordinates may be generated as follows. Suppose we begin with four points, on a non-Euclidean sur-face, arranged roughly at the corners of a square or rectangle (Fig. 6-26A). By using lasers, we can establish the geodetic lines connecting these four vertices, and then divide each of these lines into a certain number of equal partitions. This generates a set of points as shown at B. Then, from each of these points, we ascertain the geodetic line con-necting it with its correspondent, directly across the figure, as shown at C. In this example, we would

find the lines connecting X_1 with Z_1, X_2 with Z_2, X_3 with Z_3, and X_4 with Z_4; also the geodetic lines connecting Y_1 with W_1, Y_2 with W_2, Y_3 with W_3, and Y_4 with W_4. We need then bother with only sides X and Y of the figure, having coordinates (x,y) that range between zero and five (in this case). The location of a particular point P may then be found by determining the appropriate geodetic lines to sides X and Y of the figure, as shown at D.

DISTANCE THROUGH PROGRESSIVELY MORE DIMENSIONS

When space is non-Euclidean, there is no such thing as a "straight" line between two points, but only a shortest path or geodetic line, as we have seen. But given a non-Euclidean n-space, we can have paths shorter than the geodetic line when we measure

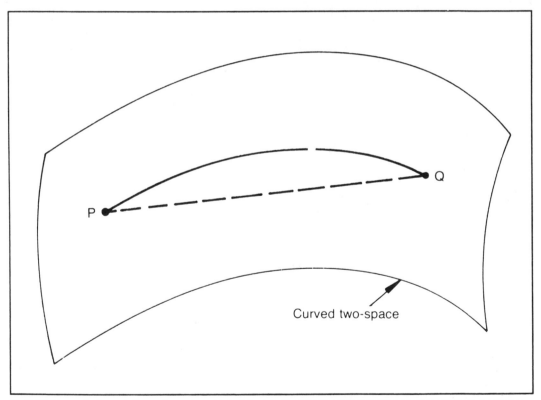

Fig.6-27. The path through three dimensions (dotted line) is shorter than that through two dimensions (solid line) in a non-Euclidean two-space. This principle holds true for any case of n dimensions.

distances through n+1-space, n+2-space, or greater spaces.

This principle is illustrated by Fig. 6-27. The dotted line shows a path through three dimensions between two points P and Q in a non-Euclidean two-space; the solid line represents a geodetic line connecting P and Q. Creatures in the two-space would measure the distance between P and Q as a certain value, but we three-space beings realize that P and Q are actually closer than that. The distance between any two points is thus smaller when measured in three-space, than it is when taken in two-space.

Our own three-space is believed by cosmologists to be non-Euclidean. If this is true, then we can conclude that any two points in our universe, separated by a given apparent distance, are actually closer to each other with respect to four-space. For we measure distances in our universe as geodetic lines, just as the two-space creatures would in Fig. 6-27. We do not yet know how to go about measuring distances through four or more dimensions, but we may someday be able to do so.

Now you must use some high-powered imagination and rather far-fetched speculation; but suppose that four-space is also non-Euclidean. Then we would measure a still smaller distance between two points in five dimensions, as compared to the measure in four-space. And if five-space is non-Euclidean too, there exists a still shorter measure in six dimensions. It can go on without end.

Although this is just a mathematical model, and we have no way of knowing (at present) whether or not it is reality, imagine that there are an infinite number of possible dimensions. It is difficult, we might argue, to provide a reason for there being any finite limit such as four, or six, or 99, or 1043 dimensions. We can escape such limitations by just postulating that there are infinitely many.

Suppose further that each continuum in n-space, regardless of the value of n, is non-Euclidean. Again, we do not know whether or not this is the case, but it is easier to imagine that a space is non-Euclidean (which can happen in so many different ways) than to suppose it is Euclidean (which can happen in only one way).

Given the above two premises, the distance between any two points P and Q gets smaller and smaller as we measure it through more and more dimensions. Through 78 dimensions, for example, we might get a distance d as the measure between P and Q; but since 78-space is non-Euclidean, we will see a shorter separation in 79-space—a distance less than d. For, in 78-space, the path d is just a geodetic path. This principle will hold no matter how many dimensions we consider.

What happens to the distance between two points P and Q as the number of dimensions increases without limit? It might converge toward some fixed value, representing the "absolute" distance between P and Q as measured through infinitely many dimensions. Or, perhaps (and this is the strangest conclusion) it vanishes to zero! In the latter case, then, the whole of all creation is just a single point in infinity-space.

Some day we may be better equipped to evaluate the truth of the two premises that lead to these rather bizarre conclusions; for now, let us return to the more accessible case of our own three-dimensional space. What has led to the conclusion that it is non-Euclidean? What properties have we observed that would make us believe it? The answer is gravitation: The presence of matter causes a distortion of distance and time. This is the basis for Einstein's general theory of relativity, which we shall now explore.

Chapter 7

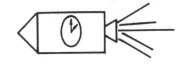

The Theory
of General Relativity

EINSTEIN'S SPECIAL PRINCIPLE OF RELATIVITY IS concerned with the motions of objects in space. It explains the observed phenomenon that the speed of light does not depend on the motion of the source, or on the motion of the observer. We have seen that this fundamental axiom leads to some strange conclusions, such as the dilation of time, the distortion of spatial measures, and the increase in inertial mass. We have seen that no object can be accelerated to the speed of light; the special theory in fact grew from the question: "What would happen to an object moving at the speed of light?"

The physics of Newton, commonly referred to as classical physics, is based on the notion of absolute space. Acceleration, according to this concept, is the result of some absolute property of the universe. In a way, this is true, as we shall see, although it is not really a property of space, but rather of what is in space.

The general theory of relativity is concerned with the structure of the universe as a whole. We will see that gravitation plays an important role in how our universe behaves. Gravity has an effect, not only on matter, but on time, space, and energy as well.

CENTRIFUGAL FORCE

We are all familiar with the "force that does not exist." A rotating or revolving mass displays an apparent outward force, sometimes called centrifugal force. This "force," while not theoretically real, does produce visible effects. (A passenger in a centrifuge at 10 g's might disagree with those who say that centrifugal force does not exist!) Rapidly rotating planets, such as Jupiter and Saturn, appear noticeably oblate because of centrifugal force. Jupiter, for example, with a rotational period of less than 10 hours as well as a largely gaseous outer structure is decidedly flattened, and this can be seen even by looking at photographs of the planet.

This outward "force" is what keeps the planets

from falling in toward the sun. It keeps the stars of our galaxy from collapsing to the center. It keeps clusters of galaxies from falling together. It also keeps a ball on a string from hitting you when you swing it around your head. This force acts on all scales, from the molecular to the cosmic. But what is responsible for it? Why can't we consider a rotating frame of reference to be stationary, and imagine the whole universe to be revolving around us? We can discover, through simple logic, what the responsible factor must be.

ALONE IN THE UNIVERSE

Imagine, for a moment, that the universe were devoid of all matter except for yourself, a length of string, and a rubber ball. Suppose you were just hovering in a void with no other things at all—no planets, stars, or galaxies. It would be quite a lonely experience, to be sure, but you might attempt to entertain yourself by tying the end of the string to the ball and swinging the ball around and around. (As a child, you no doubt did something of this sort, during a moment of sheer boredom.)

You might suppose that the ball would travel around and around in a circle, the radius of which would be determined by the length of the piece of string (Fig. 7-1). Of course, in our universe, with all its surrounding objects, there is no doubt that this

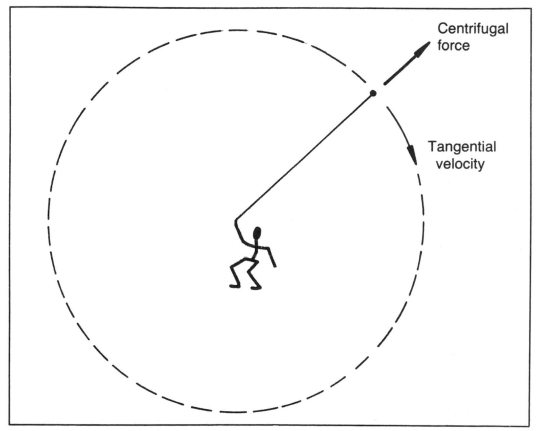

Fig. 7-1. In our universe, you can swing a ball around your head on the end of a string, and centrifugal forces will keep the string taut. But, this would not be the case in a universe with no other objects except yourself, the string and the ball, since there would be no other point of reference to ascertain rotation.

would happen. If you were to swing the object vigorously enough, it might fly off the string.

But, in a universe with no frame of reference besides the ball and yourself, it is not at all clear what would happen. How could you say with any certainty that it would be the ball, and not your own body, that was revolving? You would need some other frame of reference, an "arbiter," as it were.

A UNIVERSE OF TWO

Suppose now that yourself and a friend were suspended in a universe containing no other material objects except the ball and string. Imagine that your friend was floating about 50 feet away from you, not getting any closer or farther away (that is, the relative motion between you and him was zero).

Now, if you swing the ball around you, you can use the viewpoint of your friend as well as yourself to make the decision as to whether or not the ball is moving. Assuming your friend was not tumbling around, you would both agree that the ball was moving in a circle around your body.

But what if your friend was rotating? Then, relative to him, you would be tumbling in the opposite sense. Either point of view would be equally valid. A third arbiter would be needed to ascertain who was rotating and who was not. But even then, we could very well say that the arbiter was not an absolute reference frame at all.

In this situation, with your friend and yourself mutually tumbling but always maintaining a constant distance from each other, you could hold the ball straight out in front of you and not move it at all, and your friend would see it flying around your body along with your own rotation. Suppose the string were made of elastic material, such as rubber, so that the force of its revolution (centrifugal force) would cause the string to stretch. Your friend would then see the string stretching on account of centrifugal force; but you would not see it stretch because, to you, the ball would not be revolving. Who would be correct? Surely the string could not have two different lengths at the same time! In an otherwise void universe, we must conclude that centrifugal force cannot exist. To assume that it does leads us to a contradiction (Fig. 7-2).

It appears as if we cannot completely break free of the notion of absoluteness, in some form, in our universe. Clearly, in our universe as it exists, with billions of celestial objects, we have an absolute standard that defines rotational motion. Any rotating object experiences centrifugal force; any non-rotating object does not.

The myriad of distant celestial objects are, in fact, the only difference between our "real" universe and the "imaginary" empty ones we have just envisioned. Then it must be these masses that cause "centrifugal forces" in our space.

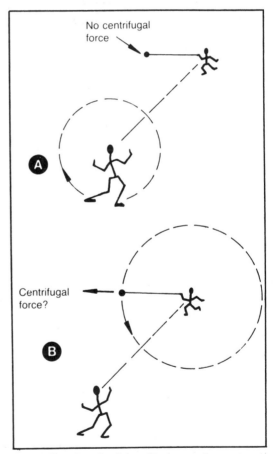

Fig. 7-2. From the point of view of the far man, the near man is rotating. But the near man thinks the far man is rotating. In a universe with no other objects, either point of view is equally valid. Then is the ball going round and round, or not?

Physicists wrestled for a long time with possible reasons for this strange effect of distant objects on all things, down to the minutest particles. It was Ernst Mach who first postulated that distant celestial objects might cause centrifugal forces. Mach's work greatly influenced Einstein. We have already seen that we cannot regard a rotating body as "fixed," because it violates the special principle of relativity by resulting in speeds greater than light for extremely distant stars and galaxies. But what is the reason for centrifugal force? This force is nothing more than a result of inertia. Inertial effects are caused by the combined gravitation of all matter in the universe.

ACCELERATION AND GRAVITATION

Centrifugal force is, of course, the result of inward acceleration of a revolving object. This acceleration is the change of velocity of such an object. As you swing a rubber ball around and around on a string, the *speed* of the ball—the number of meters per second at which it moves—may be constant, but the *velocity* is constantly changing. Velocity consists of speed and direction, and the direction is constantly changing for an object revolving in a circle. This acceleration is directly inward toward the center of a circular orbit. Inward acceleration causes an outward force, and it is this force we call "centrifugal force." The greater the acceleration, the greater this force.

The acceleration is dependent on the speed of the orbiting object, and also on the radius of the circle around which the object travels. For a constant radius of orbit, greater speeds mean the acceleration will be greater. For a constant speed, smaller orbital radius results in greater acceleration. Let the acceleration be represented by a, and the radius of the circle be represented by r; also, let T be the time required for the object to complete one revolution. Then:

$$a = \frac{4\pi^2 r}{T^2}$$

The centrifugal force, f, is the product of the mass of the revolving object and its acceleration:

$$f = ma = \frac{4\pi^2 rm}{T^2}$$

INSIDE THE BALL

Imagine that the ball at the end of the string being swung around and around is hollow. Imagine that we are somehow able to shrink down to a height of just one centimeter or so, and that we climb inside the hollow ball. Suppose then that someone takes up the string and swings it around, letting out exactly 10 meters of string (so that r = 10 meters) and orbits us with a period of T = 6.3 seconds. Assume that there is no source of gravity, such as the earth, nearby; this whole event is to take place somewhere in the far reaches of space.

According to the acceleration formula, we will experience an inward acceleration equal to:

$$a = \frac{4\pi^2 \times 10}{6.3^2} = 9.9 \text{ m/sec}^2$$

You might recognize this as approximately the same acceleration that a free-falling body attains in the gravatational proximity of our planet earth.

Now further imagine that, after having shrunk down to the tiny stature of one centimeter, we mass only 10 grams (10^{-2} kilogram). Then the force f that we experience inside the ball will be equal to:

$$f = ma = 9.9 \times 10^{-2} \text{ kgm/sec}^2$$

This is the same force that would be exerted on us by the gravity of the earth, were we to be standing on the surface at sea level. This is shown by Fig. 7-3. Of course, for any mass, the force would be the same inside the ball at the given acceleration, as compared with the force on the surface of our planet. That is to say, any object would appear to weigh the same in both situations.

How would we be able to tell, confined inside an opaque ball (to prevent us from seeing outside), whether it is revolving around on the end of a string 10 meters long with a period of 6.3 seconds, or whether the ball is sitting still somewhere on earth? The answer is, of course, that there would be no way for us to know the difference. All experiments we might conduct, confined to the interior of the

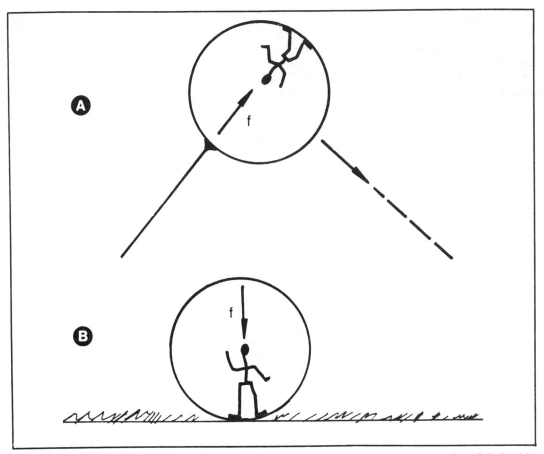

Fig.7-3. At A, a tiny man inside a hollow ball, which is swung round by a giant, experiences a certain force f. At the right acceleration, this force f is the same as if the ball were resting on the ground (at B). If the skin of the ball is opaque, so that the man inside can't see out, he will have no way to tell whether he is being swung round, or whether the ball is at rest on the surface of the earth.

ball, would yield precisely the same results in either environment.

It might be possible for the string to be longer and the period slower, resulting in an acceleration of 9.9 m/sec^2; it might also be possible for the string and the period to be shorter. In fact, there exist an infinite number of combinations of string length and orbital period that can produce an acceleration of 9.9 m/sec^2. And the ball need not necessarily be rotating; it might be a straight-line change in speed, or a combination of directional and speed changes. In any situation causing this amount of

acceleration, the environment in the ball will seem to be exactly like an at-rest condition at the earth's surface.

The force caused by velocity change is indistinguishable from gravitational force. The two forces affect everything in the same way. Einstein noticed this similarity between the force of acceleration and the force of gravity, and one of the bases of his general theory of relativity is, therefore, that the two are equivalent. That is, any acceleration resulting from a change in velocity should produce all the effects of a gravitational field, and a gravitational field should produce all the effects of a change in velocity.

DISTORTION OF TIME AND SPACE CAUSED BY ACCELERATION

Let us now return to the ball-and-string discussion, except now we will take the point of view of the person holding on to the end of the string opposite the ball. Let us say the ball has a diameter of five centimeters. The string is, as we have already mentioned, 10 meters long. How many ball diameters are in the circumference of the circle? The

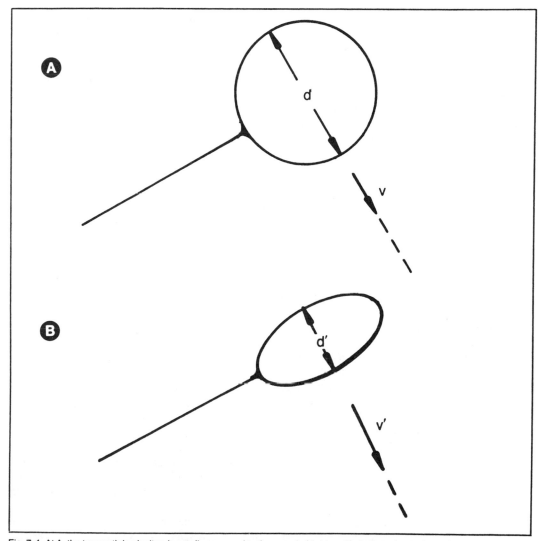

Fig. 7-4. At A, the tangential velocity v is small compared to the speed of light, so the ball appears to have the same diameter d as it does at rest. At B, the tangential velocity v′ is large, and the diameter of the ball d′ appears to be reduced in the direction of motion. This can be said to occur because of the inward acceleration, as well as because of the forward speed; the gravitational field of a planet, star, or other object has a similar effect on spatial distances.

circumference can be easily found to be about 62.8 meters; it is easy to calculate that there are 1,256 ball diameters in the circumference of the circle.

We would at first imagine that this is true regardless of how fast we swing the ball around. At slow speeds, this is true. But, as we swing the ball around faster and faster, relativistic distortion will begin to take place. Figure 7-4 illustrates the spatial change that will occur in the shape of the ball; it will appear to become oblate, or flattened, perpendicular to the string, in the direction of its motion.

If the ball attains a speed of 0.866c in its orbit around us, the diameter of the ball along the axis of its instantaneous direction will appear to be only half its at-rest value, or 2.5 centimeters, since the distortion factor at this speed is equal to 2. The string will still appear 10 meters long, however. Thus there will be twice as many ball diameters in the circumference of the circle—not 1,256, but 2,512 ball diameters. As the speed is made greater and greater, the number of ball diameters around the circle will continue to increase.

But, as seen from the interior of the ball, there will be no spatial distortion in the shape of the sphere. Everything will appear exactly normal. Riding along inside the ball, and looking out through a little hole, we would see the string as 10 meters long, just as does the observer at the orbital center. Thus we would conclude that the circumference must still be just 1,256 ball diameters. Could it be that the value of π, which we are so confident is always equal to 3.14, depends on point of view? The answer is yes. Acceleration causes distortion of the space continuum, and this can change the values of familiar Euclidean constants.

Now, suppose that we wish to measure the period of orbit. As seen from the center of the orbit, from the point of view of the person holding the end of the string, the period is:

$$T = \frac{63 \text{ m}}{0.866c} = \frac{63 \text{ m}}{2.6 \times 10^8 \text{ m/sec}}$$
$$= 2.4 \times 10^{-7} \text{ sec}$$

Figure 7-5A shows how the observer at the center of the orbit might measure the period. By observing the ball as it lines up with a distant star, and measuring the time until it lines up again with that star, the period can be determined.

But, what if we try to measure the period from the point of view of observers inside the ball? Figure 7-5B shows how the passengers in the ball might measure the period, T'. They might observe the man holding the far end of the string as he appears to line up with some distant star, and measure the time until he lines up again. The time-distortion factor at 0.866c is equal to 2; thus the passengers in the ball will see time go by only half as fast as the man at the center of the orbit senses time. The period T' will therefore be only half the period T, or 1.2×10^{-7} seconds.

We might mention that any creatures inside the ball, going around at this speed, would be lucky to survive. According to the acceleration formula, at 0.866c (2.6×10^8 m/sec) with a period $T = 2.4 \times 10^{-7}$ and r = 10 m:

$$a = \frac{4\pi^2 \times 10 \text{ m}}{(2.4 \times 10^{-7} \text{ sec})^2} = 6.8 \times 10^{15} \text{ m/sec}^2$$

This is about 6.9×10^{14}, or 690 trillion, times the intensity of the acceleration of gravity at the surface of the earth!

RESOLVING THE TWIN PARADOX

According to the theory of general relativity, it is the acceleration, as well as the speed, that causes relativistic time distortion and spatial distortion. This allows us to resolve the Twin Paradox mentioned in Chapter 3. You will recall that the idea of relative motion makes it impossible to explain changes in clock readings caused by relativistic time effects, since we are not able to say who is moving and who is not. But there is no doubt, with the rubber ball example just given, that the ball is subjected to acceleration while the man in the center is not.

Postulating that acceleration causes time and space to seem distorted, we may say with certainty that time appears to go more slowly on the ball than at the center of its orbit. The person swinging the string would see clocks on the ball going slower

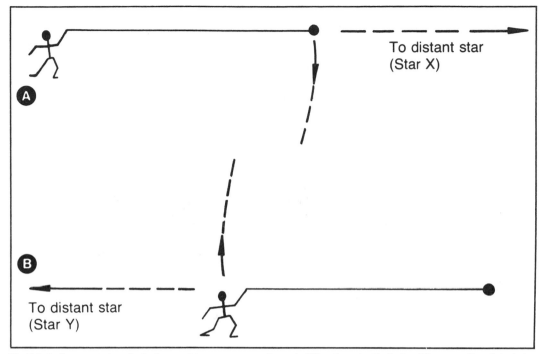

Fig. 7-5. As the man swings the ball around, he can measure the period T by observing the ball as it lines up with a distant star (A). Then, T is the time until it lines up again with that star. At B, a passenger inside the ball sees the whole universe going round; but he may measure the period by using the same method, although not necessarily the same star.

than his own, and travelers inside the ball (if they could survive the acceleration force) would see the clock of the center man running more rapidly than theirs. Upon rendezvous, the two parties would be in agreement about the difference of the clock readings.

In the case of straight-line motion rather than rotational motion, we can see that any object attaining relativistic speed will be subjected to acceleration and deceleration. It is the gravitation produced by these accelerations, and not the relative motion *per se*, that causes the disagreement in the clock readings.

Figure 7-6 illustrates gravitational effect on the rate of time progression. If two synchronized clocks are set running, as shown at A, and a source of gravitation is brought near one of them (B), the clock in the vicinity of the source of gravitation will run more slowly. At C, the source of gravitation is taken away from the proximity of the clock, and it once again runs at its original speed. But it lags the clock that is not subjected to gravitational influence. Acceleration and deceleration on a space vessel are equivalent to the presence of a gravitational field, and hence they cause the clock on board the vessel to go more slowly than when accelerations are not present. This causes a time lag, and consequent disagreement in later clock readings.

THE ROTATIONAL ACCELERATION FIELD

We are all familiar with the inverse-square gravitational field that exists around all material objects. The force on a given mass becomes smaller as the distance from the source gets greater. The function is given by the equation:

$$f = ma = \frac{mk}{r^2}$$

125

where m is the mass of the object being influenced, r is the distance from the source of gravitation, and k is a constant that depends on the mass of the gravitational source. This "inverse square" relation is shown by Fig. 7-7A.

Suppose that the string in the preceding orbiting-ball example is replaced by a stiff rod of some sort, that remains straight under all conditions. Suppose the rod is swung around and extends an indefinite distance from the center of rotation. Further suppose that we are able to move freely along this rod, varying the radius r at will.

When r = 0, at the center rotation, the acceleration will be zero. As we get farther and farther away from the center, the acceleration will increase. When r = 1 m, we will have $a = 4\pi^2/T^2$ m/sec²; when r = 2 m, we will have $a = 8\pi^2/T^2$ m/sec². In general, the acceleration, a, will increase linearly with the distance, r, from the center of rotation. For a given mass, then, the force f exerted on it is a linear function of the distance from the center of revolution. Figure 7-7B is a graph of this relation. It is in contrast to the familiar gravitational-field function; in the rotating acceleration field, the force *increases* with distance, and this increase is without limit.

This outward centrifugal force is equivalent to a gravitational field that becomes more and more

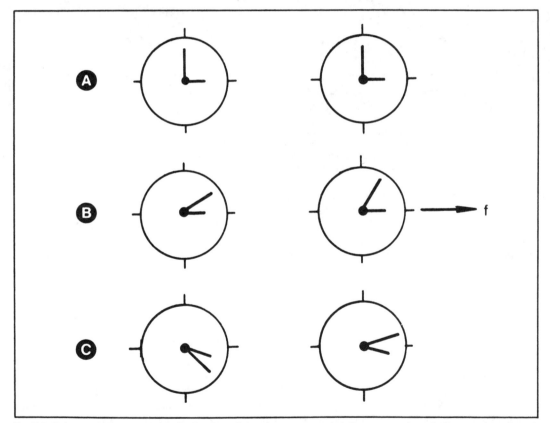

Fig. 7-6. At A, two clocks are synchronized and running at the same speed. At B, a source of gravitation, exerting a force f, reduces the speed of one clock. At C, the source of gravitation has been removed, and the second clock returns to its previous rate. But it is now lagging the other clock by an amount that depends on the intensity of the gravitational field and the length of time it was present.

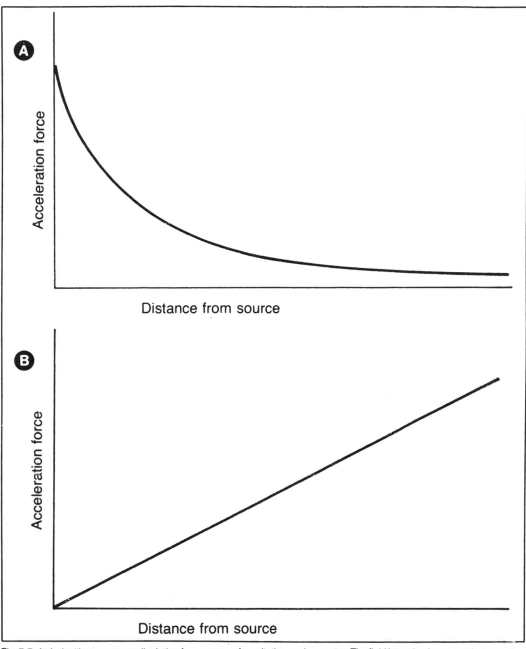

Fig. 7-7. At A, the "inverse square" relation for a source of gravitation such as a star. The field intensity decreases in proportion to the square of the distance from the source. At B, the relation for a rotating frame of reference, such as a ball and string, or a rigid rod. The intensity of the force increases in direct proportion to the distance from the center of rotation.

intense as we get farther from the "source," which is the end of the rod at which the rotation is generated. But, as we get to great distances from the center of rotation, what will happen? Eventually, the speed of revolution will reach the speed of light, since the rod is absolutely rigid, will it not? For example, if $T = 1$ sec and $r = 4.8 \times 10^7$ m, the speed of revolution becomes 3×10^8 m/sec, or the speed of light, does it not?

Obviously, the answer must be no. Something must happen to prevent this. One might be tempted to imagine that the stiff rod would have to be limited in its length if it were to remain straight, and that this length would depend on the period of revolu-

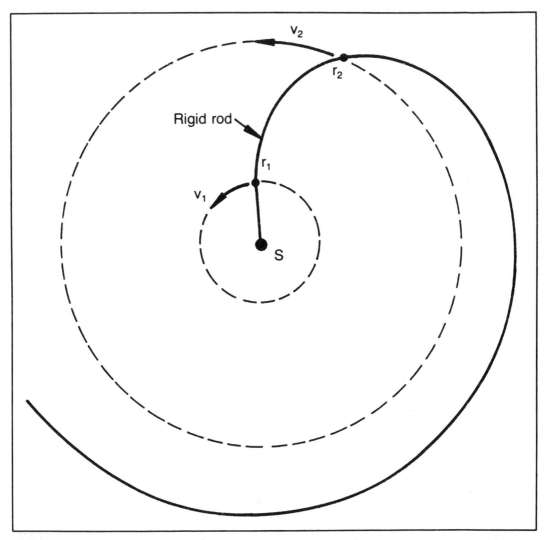

Fig. 7-8. Distortion of a "rigid" rod as it rotates around a center S of rotation. At radius r_1, the tangential velocity is V_1. Although the tangential velocity would seem to exceed the speed of light at sufficient distance r_2, this is contrary to the laws of special relativity. Thus $v_2 < c$, and the rod is curved.

tion, T. If T = 1 sec, perhaps the rod can only be 4.8 × 10^7 m in length, and beyond that distance, it cannot exist! Doubling the angular rate of rotation, reducing the period to T = 0.5 sec, might then reduce the rigid rod to a length of 2.4 × 10^7 m. Faster and faster angular velocities might make the rod shorter and shorter until, at sufficiently high rates, it might be only a few centimeters long! But this is of course an absurd explanation.

A more acceptable means of explaining this apparent paradox can be realized by asking ourselves the question: "What makes us believe the rod is straight?" The answer is that it *looks* straight; a light beam sent out along the rod follows the rod. Visually, the rod appears to be straight. But, this assumes that the rays of light themselves move in straight lines, and we have seen that this need not necessarily be the case. Light beams travel in geodetic paths, but these paths may be curved.

The rod, then, might occupy a position that does not describe a perfectly straight line in space, but instead, a geodetic line that represents the straightest possible existence given the parameters of the surrounding universe. Figure 7-8 shows this. Near the center of rotation, where the tangential speed is not very great, the rod is nearly straight. But farther and farther from the center of rotation, as the speed increases, the rod is bent into a spiral shape, as viewed from a reference point far away from the rod itself. Then, the speed of any point on the rod need never exceed the speed of light.

ROTATING COORDINATES

The example just described gives us a way to establish equivalence among reference frames that are rotating with respect to each other. The theory of special relativity shows how inertial frames of reference are equivalent; it is simple to establish a relation between any two coordinate systems that move with constant velocity. The theory of general relativity carries the propositions of special relativity further: *Any* two points of view are valid observation points in the universe; we can find an equivalence relation between frames even if they are rotating or accelerating.

It is not immediately clear how a rotating coordinate system can be used as a reference frame free from contradiction. The rotating earth is a good example. From the vantage point of an observer at the North Pole, all the objects in the universe seem to go round and round. The sun completes one revolution in exactly 24 hours; the moon takes nearly 25 hours; the distant stars take a little less than 24 hours. But, at distances such as those from the earth to the stars outside the solar system, extreme speeds seem to occur in their daily trip around us—speeds much in excess of the speed of light. A star just 10 light years away, and in line with the celestial equator, will appear to traverse a circumference of 63 light years in only a day. This is, apparently, 23,000 times the speed of light!

This is a hasty conclusion, however, for we have not established a coordinate system for the rotating point of view of an observer standing at the North Pole on the earth. Certainly we cannot deny that the star is 10 light years away, since we can use parallax techniques to verify this. And it obviously does go once around in 24 hours; this can be seen just by looking at a clock. The apparent contradiction occurs because we subconsciously assume that a coordinate system for this rotating reference frame must have straight-line axes. From the previous chapters, we have seen that the "straightest" lines we can ever get are geodetics. A beam of light follows a geodetic line through space.

Figure 7-9 shows a polar coordinate system according to a man standing at the North Pole of the planet earth, *as he would imagine it*. Each succeeding concentric circle represents one light hour. The radial line L represents the zero angle. The observer may thus determine the position of a point in space by means of an angle θ, measured counterclockwise from the radial line L, and by measuring the distance r in light hours. (Actually, this coordinate system is restricted to a plane, parallel to the celestial equator and passing through the North Pole. We have dimensionally reduced the coordinate system for ease of illustration.)

There is a serious flaw in this system, as it is shown in Fig. 7-9. How do we know that L is a geodetic line? It is, in fact, *not* a geodetic line.

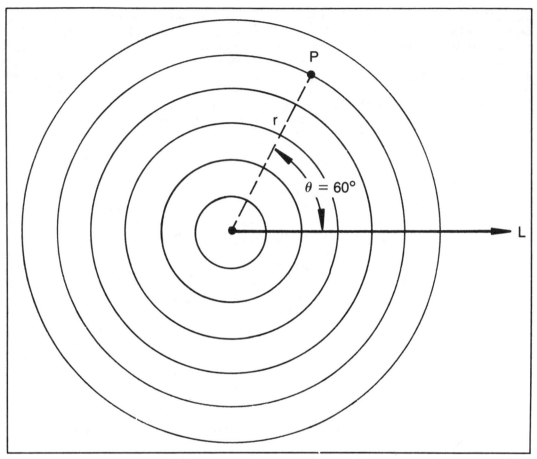

Fig. 7-9. The polar coordinate system, as imagined by an observer standing at the North Pole of the earth. A point in space is determined by measuring the angle θ counterclockwise from a reference line L, and by measuring the radial distance r. Here, each concentric circle represents one light hour. This puts P at the distance r = 5 light hours, and θ = 60 degrees.

All beams of light follow geodetic paths through space, and we may therefore use a single photon to show the position of a geodetic line in space. Figure 7-10 illustrates the path of a photon emitted in the coordinate plane—that is, initially along line L in Fig. 7-9—as observed from high above the North Pole, but from a *non-rotating* point of view. The earth appears to rotate counterclockwise, completing one turn of 360 degrees in about 24 hours.

At A, the photon is just leaving the earth, and travels in a line that appears straight. The photon

continues along this straight line, and after three hours have passed (B), the earth has turned 45 degrees and the photon has reached a distance of three light hours from the planet. The photon is therefore not over the same longitude as previously. At C, after a full day, the earth has made one complete turn, and the photon is again at the same longitude as it was when it first left the planet. It is now 24 light hours distant. As the days go by, the earth keeps on rotating and the photon keeps on going in a straight line.

Now, let us set ourselves, as we hover high

above the pole, in rotational motion, so that we no longer see the earth to be going around. Instead, the rest of the universe goes around us. Figure 7-11 shows how we will observe the path of the photon from this point of view. At A, the photon is just leaving the planet, and everything appears the same as from the non-rotating point of view. But this sameness does not last for long. By the time three hours have passed, the universe has turned 45 degrees in a clockwise direction (B), and so has the path of the photon. We see the photon describe a curved route as it moves away from the earth. After a full day has elapsed, the photon will appear 24 light hours from the earth, but it will seem to have made a spiral around the planet, as shown at C. The path of the photon cannot possibly be a straight line from this reference frame. It must be in only one place at one time, and its position is always the

same according to either the rotating or non-rotating vantage point, with respect to the universe at large.

As the days go by, and the earth rotates many times on its axis, the photon will appear to describe a spiral that gets wider and wider. The path of the photon will approach a perfect circle as its distance gets extreme. This spiral, according to the rotating reference frame, is a geodetic line because, by definition, a photon must travel a geodetic path through space.

It is a simple mental exercise to put the light-hour circles in Fig. 7-11. This is shown in Fig. 7-12; we now have a coordinate system for the rotating reference frame of the observer at the North Pole. The radial line L', for $\theta = 0$, is a spiral rather than a straight line.

Of course, the speed of the photon is always

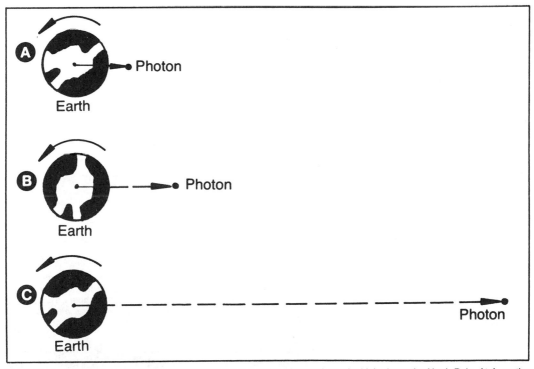

Fig. 7-10. A photon leaves the earth, viewed from a non-rotating observation point high above the North Pole. At A, as the photon is just leaving; at B, after three hours; at C, after one full day. The photon appears to describe a straight line through space. (Distances in this drawing are not to scale.)

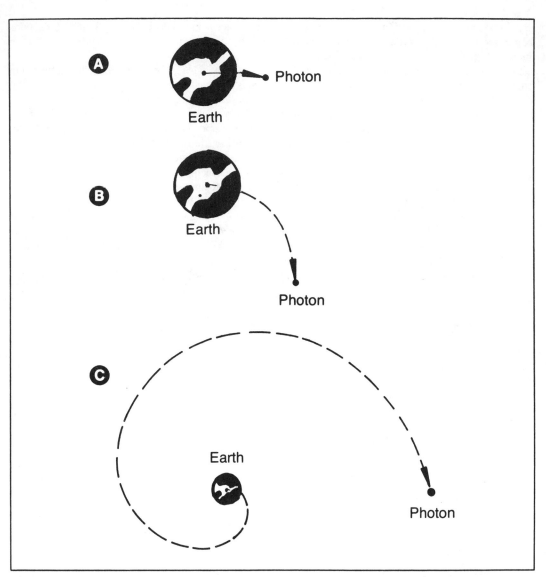

Fig. 7-11. The same photon as shown in Fig.7-10, but from a rotating observation point high above the North Pole. The observer is rotating along with the rotation of the planet. At A, the photon just leaves earth. At B, after three hours, it is apparent that it is not traveling in a straight line. At C, after one full day, the photon has gone once around the planet in a spiral path. (Distances are not to scale.)

equal to the speed of light; this is trivial. But, according to the rotating point of view, the photon appears to move faster and faster as it gets farther and farther from the earth. Each hour, the photon reaches the next "light-hour" circle in the coordinate system; but each hour, it must travel a greater distance to get to that circle. This distance seems to increase indefinitely.

132

Very distant stars may seem to travel many times the speed of light, according to the rotating point of view. But, distances appear greatly distorted, since the coordinate system is twisted many times around itself. At stellar distances, the line L' is nearly a perfect circle, as shown by Fig. 7-13. But, the line L' is never moving at more than the speed of light with respect to any star. The speed of a star relative to L' approaches c as the distance increases. Thus the stars are really not going around at impossible speeds. The angular distances are greatly distorted because of distortion in space itself.

ALL POINTS OF VIEW ARE EQUALLY VALID

The rotating coordinate system of Fig. 7-12 hardly looks like the non-rotating one of Fig. 7-9, but they both represent the same universe. It is possible to obtain a one-to-one correspondence, or homeomorphism, between the two systems of coordinates, and this enables us to define all events in one system in terms of the other. Thus the two points of

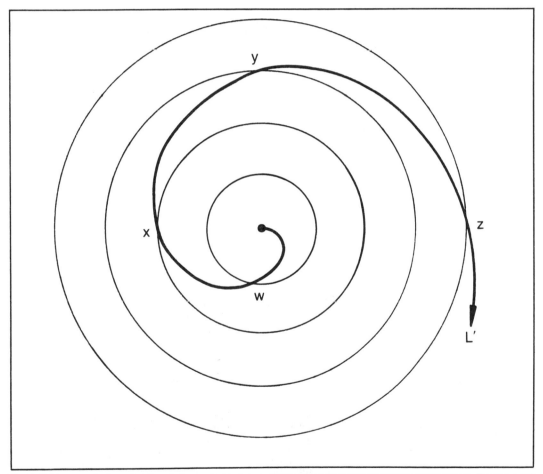

Fig. 7-12. The coordinate system for rotating earth. Each succeeding concentric circle represents six light hours. The line $\theta = 0$ is shown by L'. In this system, distance wx, xy and yz are each the same—six light hours. Thus we see that the distance becomes distorted as we get farther from the earth, and this distortion increases with increasing radius r.

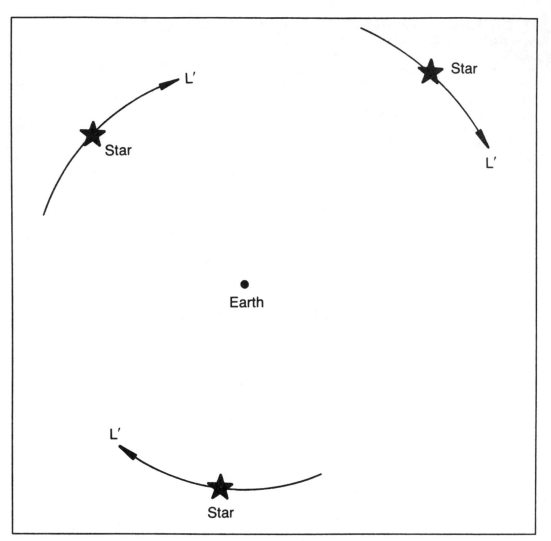

Fig. 7-13. At stellar distances, the $\theta = 0$ line, L', is almost a perfect circle. One complete circumference around the earth is never more than 24 light hours, even though it would seem to be many thousands or millions of light years. Consequently, even the most distant galaxies move at less than the speed of light according to the rotating point of view.

view, rotating and non-rotating, are equally true pictures of the universe. No contradiction need result in the laws of physics as seen from any reference frame.

No matter what the point of view, a coordinate system can always be found that enables us to observe the behavior of the universe according to the same set of physical laws. All that is necessary is that a homeomorphism exist between the system of coordinates we use, and some other established set of coordinates. Nor does it matter what that established set is. It may correspond, itself, to any point of view within the universe. Let us observe some examples of homeomorphisms.

ONE-TO-ONE CORRESPONDENCES

Imagine a universe of three spatial dimensions, such as our own, defined by means of a Cartesian system of coordinates. We might put the origin of this coordinate system anywhere, and it might be moving at any speed. It could be situated on Jupiter as it revolves around the sun and rotates on its axis once every ten hours; it might be on an asteroid tumbling in orbit in the solar system.

Suppose that Jim and Joe are aboard two space vessels, as shown in Fig. 7-14. They may each set up their own Cartesian coordinate system, and place themselves at the origin (0,0,0). The units they use might be of different lengths; Jim might use one foot and Joe might use one meter. According to either observer, the position of the other appears to constantly change within the coordinate system, since the two ships are moving with respect to each other. At some particular instant of time, Joe might see Jim at position (x_0, y_0, z_0) by his own system, but this value will constantly change. The same is true for the way Jim sees Joe. At some moment, Joe might be at coordinates (u_0, v_0, w_0), but this value constantly changes as time passes.

No matter what the relative motion between the two ships, one thing is always true: There is a one-to-one correspondence between the points in Jim's space and those in Joe's space. This means that every point in Jim's system can be paired off with one, but only one, point in Joe's universe, and vice-versa. Never does Jim have a point that cannot be assigned a mate in Joe's system. Never does one astronaut have two points that can be assigned to a single point in the other's universe. At all times, a perfect one-to-one correspondence can be found. Thus, Jim's universe is homeomorphic with Joe's.

It makes no difference whether Jim's ship is tumbling and Joe's ship is in orbit about Mercury; it does not matter if Jim is accelerating toward the center of the galaxy while Joe sits on the moon eating a sandwich. It is always possible to find a one-to-one correspondence between their universes. If this homeomorphism did not exist, then the two universes would not be equivalent; Jim's point of view would somehow be different from Joe's.

How might two universes not be homeomorphic? This can happen if space becomes "ripped" or "creased." Figure 7-15 shows two plane universes that are not homeomorphic because one of them has a "tear" in it. Figure 7-16 shows how a plane universe might be folded over on itself ("creased") in

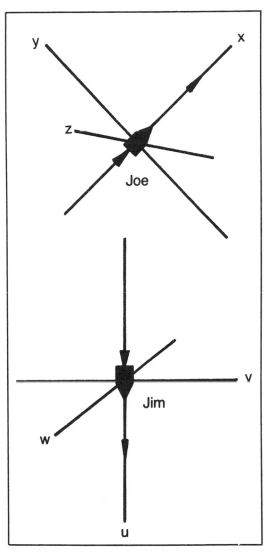

Fig. 7-14. Two astronauts, Jim and Joe, travel in space and set up their own Cartesian coordinate systems. Joe uses the variables x, y and z; Jim uses the variables u, v and w.

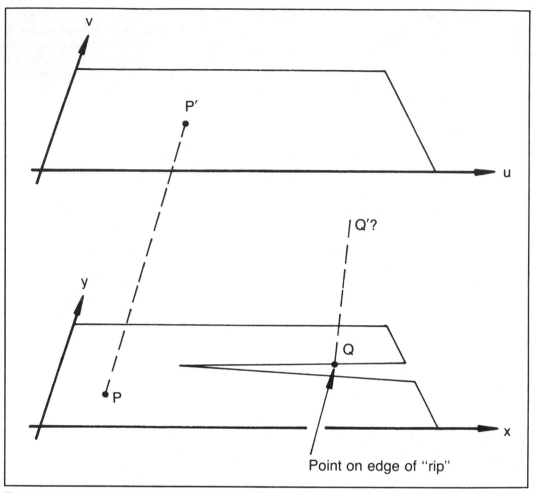

Fig. 7-15. Two plane universes, one of which has a "rip." A point P in the "ripped" universe has a mate in the other space, but a point Q on the edge of the "rip" cannot be paired off with any point in the other universe. If we do assign a point Q' to point Q, then some other points in the homogeneous universe will be left wanting. These spaces are not homeomorphic.

such a way that a one-to-one correspondence cannot be found between it and a flat plane.

Do such things ever happen to our own universe? Are there any reference frames in which many points become compressed into a single point, or from which some points disappear altogether? The answer is no, insofar as motion and acceleration are concerned. No matter what Jim or Joe do, no matter what contortions they may put themselves through in an effort to "rip" or "crease"

the observed space according to the other astronaut, they cannot succeed at it. But, certain gravitational fields can create "holes" in the continuum. We shall have more to say about these fields, known as black holes, in the next chapter.

In general, then, for any two reference frames in the universe, it is always possible to find a homeomorphism between the two resulting universes. By means of a suitable transformation, all events according to one viewpoint can be defined

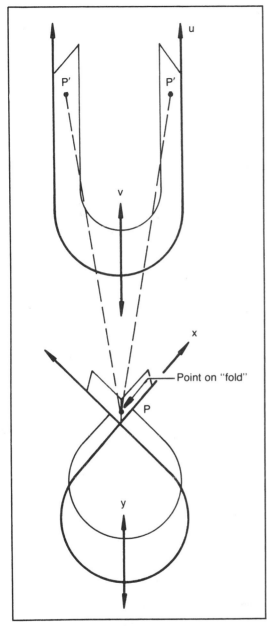

Fig. 7-16. Two plane universes, one of which has a "crease." A point P on the "crease'" has ambiguous mates in the other space, which is greatly distorted from the Euclidean, but not to the extent that it is "creased." These spaces are not homeomorphic.

according to the other viewpoint, and all the laws of physics remain intact. The transformation may be complex, but it always exists. This is an important result of the theory of general relativity. Space cannot be "ripped" or "creased" by relative motion.

We are now able to see how the effect of spatial distortion makes it impossible for any object to attain the speed of light. If this were possible, then a hollow sphere would become distorted to such an extent that it attained a disk shape. But that would destroy the homeomorphism between a point of view inside the sphere, and a point of view outside and moving more slowly. A single point on the disk would sometimes correspond to just one point on the sphere, but this would only be true if the point on the disk were exactly at the edge. Any point inside the disk would have two mates on the sphere (Fig. 7-17). No matter how we might try to rearrange the points in an effort to find a homeomorphism, we could not succeed. Such a state of affairs violates the general theory of relativity.

ACCELERATION DISTORTS SPACE

When two different coordinate systems are stationary with respect to each other, or when their relative velocity is constant, the transformation between the two is fairly simple. In the example of Fig. 7-14, as long as Jim and Joe have a constant relative velocity, then straight lines in Jim's universe will also be straight at Joe's universe, and vice-versa. Objects may appear spatially distorted, but a straight rod in Jim's ship will look straight to Joe, too, no matter how the rod is oriented, although its length may not appear the same. When all straight lines in one coordinate system look straight according to another coordinate setup, we say that the two systems are related by a linear transformation. Otherwise the transformation is nonlinear.

Now, imagine that the velocity between the vessels is not constant. Then, straight lines in one universe may not always appear straight according to the other reference frame. Jim's straight rod might look bent to Joe, and Joe's straight rod might look bent to Jim. In this case, the transformation is nonlinear, but there is still a homeomorphism between the two spaces. A one-to-one correspon-

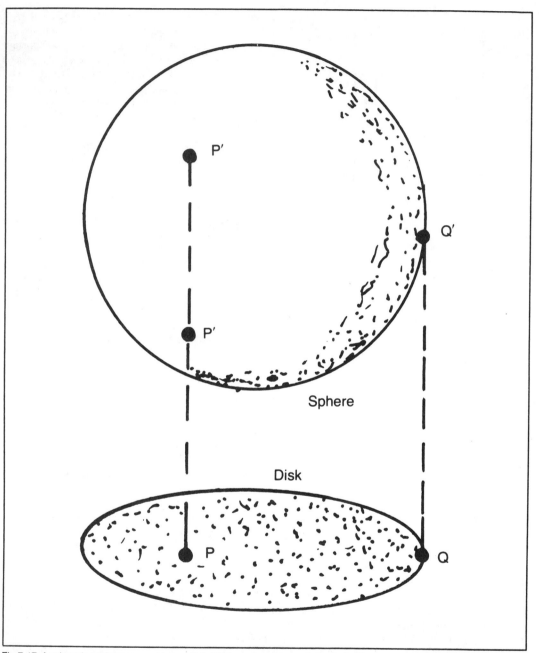

Fig.7-17. A sphere is not homeomorphic to a disk. While some points can be paired off in one-to-one correspondence, such as Q and Q', other points on the disk will correspond to two points on the sphere, such as the interior point P, which has ambiguous mates.

dence still exists between the points of one universe and the points of the other.

Figure 7-18 is a "stop-action" illustration of Jim's ship as it might appear as it moves past us at constant velocity. There are four drawings of the ship, depicting its position at times t = 0, t = 1, t = 2 and t = 3, in arbitrary units. (It does not matter what the actual length of the time unit is, as long as it remains the same.)

Let us suppose that Jim shoots a single photon of light outward from one wall of his ship toward the other wall, directly across (laterally). It takes, by our reckoning, three time units for the photon to get once across the ship. We see the path of the photon as a straight line. Jim sees it as a straight line, too, although a much shorter line, because of the relativistic speed at which his ship is traveling. Although there is a great difference in the path length as seen from the two points of view, the transformation between universes is linear: The

photon goes in a straight line according to both reference frames.

But, now suppose that Jim's ship is increasing in forward speed. Figure 7-19 illustrates this situation. We see the vessel, as before, at times t = 0, t = 1, t = 2 and t = 3; but because of acceleration, the "stop-action" images are spaced at ever-increasing intervals.

From our external point of view, the photon appears to travel in a straight line, and in fact in exactly the same position as it would if the ship were not accelerating. This is to be expected, since the speed of light does not depend on the speed of its source, but only on the viewpoint of the observer, who always finds it to be the same. But, according to Jim, the photon is falling astern. He will see the photon follow a curved path. As we see things, this is just a natural result of changing forward speed of Jim's vessel; Jim is running away from the photon! But to Jim, the curvature of the

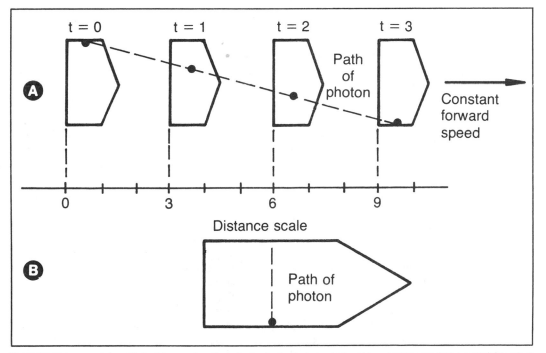

Fig. 7-18. At A, we watch as Jim's ship goes by at constant velocity. A photon in the ship seems to us to follow a straight path. It seems that way to Jim, too, as shown at B.

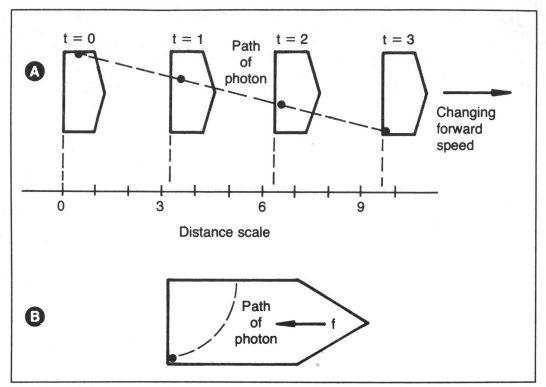

Fig. 7-19. At A, Jim's ship is now accelerating, as can be seen by observing the distance scale markings and comparing with those in Fig. 7-18. The path of the photon is just the same as it would be if the ship were not changing speed, as seen from our point of view. But Jim sees things differently; the photon seems to be "falling" in the direction of the acceleration force, f.

photon's path is very real, and he must conclude that space is curved in his environment. Obviously the transformation between our set of coordinates and Jim's is nonlinear.

If Jim's ship were decelerating rather than accelerating—that is, if the ship were constantly slowing down—the photon would seem to Jim to veer toward the front of his vessel. Sideways acceleration would cause the photon to follow a straight path, just as if the velocity were constant; but the energy contained in the photon would change. This would be manifested as a change of mass, not speed, since the speed of light is an absolute invariant. If the acceleration force f were in the direction of the photon's travel, the mass would increase; if f were against the direction of the photon's travel, the mass would decrease.

DEFLECTION OF A PHOTON WITH CIRCULAR MOTION

Let us return to the example of acceleration caused by a change in direction, rather than a change in speed. This might be depicted by the previous example of the little man traveling around and around inside a hollow ball at the end of a string. Suppose this little man carries a flashlight, and suppose he shines the flashlight in the direction of the tangential motion of the ball (Fig. 7-20). We may view this event from a vantage point high above the plane of the orbit of the ball, as shown in A; or we may be traveling along with the little man, next to the ball, so that we can look inside and see his environment from his point of view (B).

As we trace the path of a photon from the little man's flashlight, the photon appears to describe a straight line as seen from the viewpoint at A, as long as we ourselves are not rotating relative to the distant masses in the universe, and as long as the rotational center of the ball's orbit is not accelerating. In such a case, geodetic lines in the universe appear straight. But, to the little man, provided his speed is sufficient compared to the speed of light, things look much different. To him, if his orbital speed is large enough, the photon will seem to curve downward in the direction of the centrifugal force. The greater the force, the greater the amount of curvature to the path of the photon.

We can see, from an external viewpoint such as that at A, clearly why this is so. We would regard it as an optical illusion; the ball is apparently being "pulled" inward from the path of the photon. But, according to the little man inside the ball, who knows only what he sees, the curvature is very real, and he must therefore ascribe it to the acceleration force. He must also conclude that at least some geodetic lines in his universe are curved lines. Only

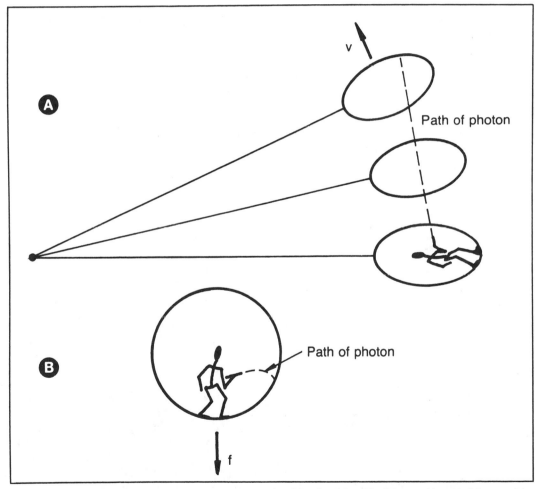

Fig. 7-20. At A, a revolving rubber ball as seen from a non-accelerating point of view. The little man shines his light toward the shell of the hollow ball, in the direction of its tangential movement of velocity v. The photon travels straight, but the ball seems to pull away toward the center of revolution. At B, the little man sees this as a curved path for the light.

if he were to shine the light exactly toward or away from the center of revolution would he find a straight geodetic line.

THE PRINCIPLE OF EQUIVALENCE

Suppose that both Jim, in the example of Fig. 7-19, and the little man, in the example of Fig. 7-20, experience an acceleration of about 9.9 m/sec^2. They will recognize this acceleration as "one gravity," the same intensity as that on the surface of the earth. (With an acceleration of only one gravity, the degree of geodetic-line curvature is extremely small, but it might be observed with sufficiently sensitive observation devices.)

If Jim's ship has opaque walls, so that he has no way to observe his surroundings, he might feel quite at home. He would feel no different than if his vessel were standing on end somewhere on his home planet. The little man inside the hollow ball would also feel at home (assuming his planet has the same gravitational field intensity as the earth). If the ball were opaque, he might well conclude that the ball is at rest on the surface of his planet, or the earth.

Are there any experiments that Jim, or the little man, might perform to determine whether they are accelerating in the far reaches of space or whether they are at rest on a planet? Might some kind of instrument, confined entirely to the interior of the ship or the ball, indicate the actual situation? The answer is no; we do not know of any way to tell the difference between a gravitational field such as is produced by a nearby planet or star, and the acceleration resulting from a change in velocity. By all the laws of physics, when we are confined to a small place, we know of no means by which we can differentiate between the two. The two forces are therefore equivalent.

This is the "principle of equivalence," and its consequences are far-reaching indeed. All phenomena that take place in a "true" gravitational field will also be observed in an "artificial" field; the converse is also true. This applies equally well to photons of radiant energy, a flying piece of rock, or a scale.

Rays of light, then, should describe curved paths in the presence of a gravitational field near a planet or star. At the surface of the earth, for example, a beam of light sent parallel to the surface should not describe a perfectly straight line through space, but rather a path that is slightly bent downward (Fig. 7-21A). This effect is very, very small in the earth's gravitational field, but it nevertheless does occur. A ray of light passing close to the earth should be bent slightly in the direction of the planet (Fig. 7-21B).

These effects become more significant as the gravitational field becomes more intense. The path of a photon would be bent much more, for example, in the vicinity of the sun. If a source of gravitation is massive enough and dense enough to produce a sufficiently intense gravitational acceleration, a photon might actually be returned toward its source (Fig. 7-21C)! In theory, it is possible that gravity might get so powerful that a photon could be captured and never released. Then we would have that situation about which so much fascination revolves: a "black hole," where things can get in but cannot get out.

According to the gravitational physics of Newton, the deflection of a ray of light passing near a gravitational mass is simply due to the fact that photons are projectiles with a certain mass and speed. Then, the gravitational attraction of a nearby celestial object should cause a change in the path of the photon, in the same way that a space vessel would be affected (Fig. 7-22). In fact, it is possible to calculate the degree of deflection of a light beam in a gravitational field, according to this Newtonian model. But such calculations have not produced accurate results. The values obtained according to this model are approximately half the values derived from actual observation. The relativistic model, however, predicts the curvature of light passing near the sun to a high accuracy, as we shall shortly see, and there are other observed phenomena that are not accounted for according to Newton's theory but are predicted by Einstein's general theory of relativity.

In Einstein's model, gravitation is simply a curvature of space, manifested by the bending of geodetic lines. A light beam passing between two

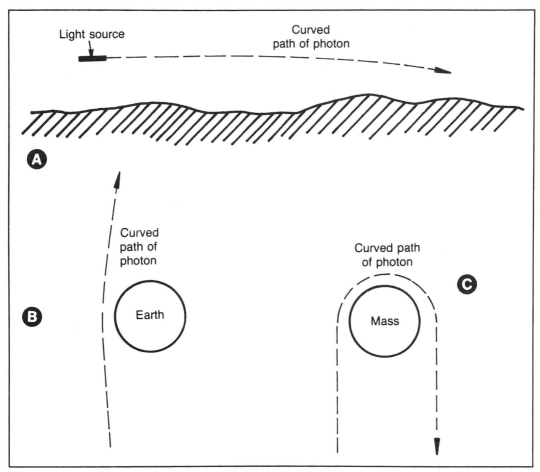

Fig. 7-21. At A, the gravity of the earth should cause a beam of light to appear bent. At B, a photon passing near the earth is deflected slightly. At C, a photon passing near a very dense celestial mass may be substantially deflected.

points is "bent" by a nearby mass because space itself is bent. Thus the straightest possible path is curved; there are *no* straight lines. This model is illustrated by Fig. 7-23.

Imagine a sheet of thin, very elastic rubber, attached to the walls of a small room at a uniform height above the floor. Suppose this sheet is stretched tightly so that it is almost perfectly smooth and flat. This sheet of rubber represents a model of two-space in the absence of gravitational forces; geodetic lines within this universe are almost perfectly straight, since the space is nearly

Euclidean. The only deviation from a perfectly Euclidean continuum results from the tiny amount of sag caused by gravity.

A creature living in this two-dimensional world would observe geodetics as straight lines. No observations he might make would lead him to believe that his universe deviated, in even the slightest measure, from the Euclidean. But, now imagine that a heavy coin is dropped onto the rubber sheet. This will cause a change in the curvature of the sheet, especially in the immediate vicinity of the coin, but to some extent everywhere. There will be substantial curvature of geodetic lines near the coin. Now, the little two-dimensional being, himself distorted right along with the space in the region near the coin, will not be aware of any change unless he is very observant. But, if he looks very closely at the position of a distant point on the rubber sheet before and after the introduction of the coin, he will note a slight change. If his power of intuition is great enough, he might be able to figure out the actual situation. But, he may only be mystified; he might conclude that the presence of the coin has produced a defect in his instruments.

LIGHT FROM THE STARS

Einstein's idea that gravitational fields are actually a "bent" space leads to certain predictions about the behavior of light beams, as we have just seen. The theory also leads to predictions concerning the orbits of celestial objects, such as the motions of the planets round the sun. The theory of general relativity also introduces the idea that there might be such things as "gravity waves"—ripples in space, like the ripples caused by a stone impacting on a water pond. Scientists have observed all these things, and the results they have obtained agree, to

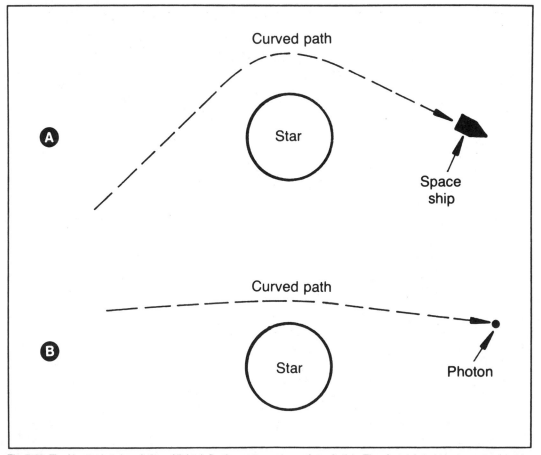

Fig. 7-22. The Newtonian description of light deflection near a source of gravitation. The photon is bent just as a space ship, coasting at constant speed, would be deflected (A); the photon is of course deflected to a much lesser extent because of its great speed (B).

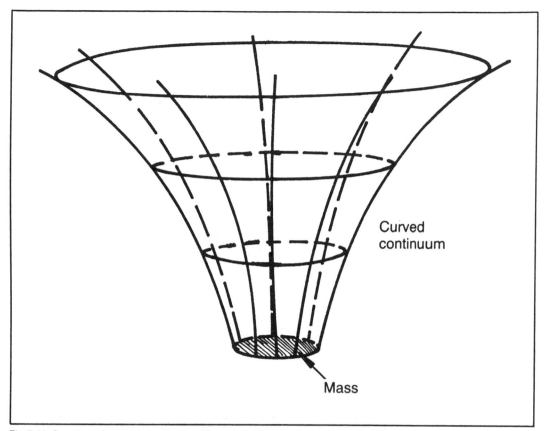

Curved
continuum

Mass

Fig. 7-23. Curvature of a two-dimensional continuum in the locality of a gravitational mass. Straight lines do not exist. A polar coordinate system is roughly shown.

a reasonable extent, with the predictions made by Einstein. No observations have yet been made to conclusively refute the theory.

The gravitational field of the earth is not strong enough to produce noticeable distortion of space, even near the horizon. The gravitational fields of the moon and planets are also too weak to be used for experiments to detect the curvature of light arriving from distant stars. But, the sun has a field of adequate intensity. Figure 7-24 shows how an experiment may be carried out to observe the bending of light rays from distant stars. This experiment was actually carried out in the early part of this century to verify the general theory of relativity.

In Fig. 7-24A, the position of a star is carefully noted. The star must be on, or very close to, the plane of the ecliptic, so that it will be eclipsed by the sun. The observation is first made in the absence of the sun. As the year progresses, and the earth revolves around the sun, the star will eventually be eclipsed by the sun's disk. Just before the eclipse of the star occurs, the star's position is again noted (B). If the path of the photons from the star to the earth is altered by the gravitational field of the sun, then we would note a change in the position of the star as compared with its position earlier.

This experiment was conducted on May 29, 1919, by two British astronomers during a solar eclipse. They took several photographs, and in

145

November of that same year, they announced that the predictions of the general theory of relativity had been confirmed. A change in the positions of stars near the sun was indeed seen (Fig. 7-25); at a time when the sun should have eclipsed a star, it was still visible near the edge of the sun's disk.

The amount of change in the positions of stars near the sun was very close to the function predicted by the general theory of relativity. Scientists who supported this theory considered the experiment a great triumph. But other evidence had to be found to provide more conclusive proof of the theory. Newton's theory also would predict the same effect, and while the deviation in stellar positions predicted by Newton was only half the observed amount, and only half the amount predicted by the general theory of relativity, the error could be traced to a simple miscalculation by Newton concerning the intensity of the sun's gravitational field. Some effect had to be observed that would agree with the general theory of relativity, but was entirely neglected by the physics of Newton. The orbit of the planet Mercury proved to be the answer to this search.

THE PERIHELION OF MERCURY

The orbits of the planets around the sun, and in fact the orbits of most satellites, are almost perfect circles. But there is always some deviation; the orbits are usually somewhat "lopsided." At a certain point in the orbit of a planet, its distance from the sun is at a minimum, and the direction of this *perihelion* can be very accurately determined with respect to the distant cosmos (Fig. 7-26).

The perihelion of each planet changes very slowly. On each revolution, the direction of the perihelion is just a tiny bit different. For the outer planets, this change in the perihelion, as predicted by Einstein's theory, is too small to be measured using presently available instruments. However, for the inner three planets—Mercury, Venus, and Earth—it is measurable. Mercury, whose proxim-

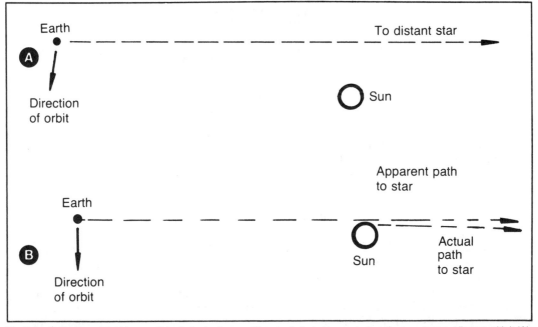

Fig. 7-24. Deflection of light from a distant star by the sun. The star is first observed when the sun does not line up with it (A); then, just before the sun passes in front of the star, its position is noted again (B). Because of the deflection of the light arriving from the star, its position will appear to change.

146

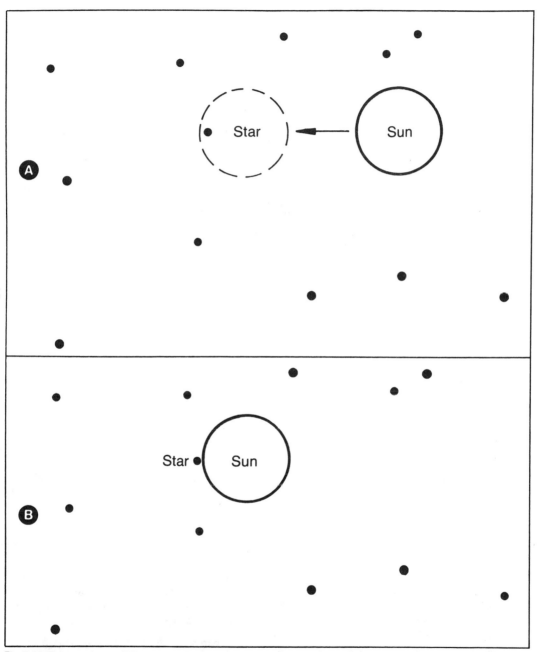

Fig. 7-25. The experiment of Fig.7-24, as observed from the earth. At A, the star is not near the sun. The dotted circle shows where the sun will soon pass. At B, the sun has arrived at the predicted position, but the star is not yet eclipsed, as we would have guessed from the prediction at A.

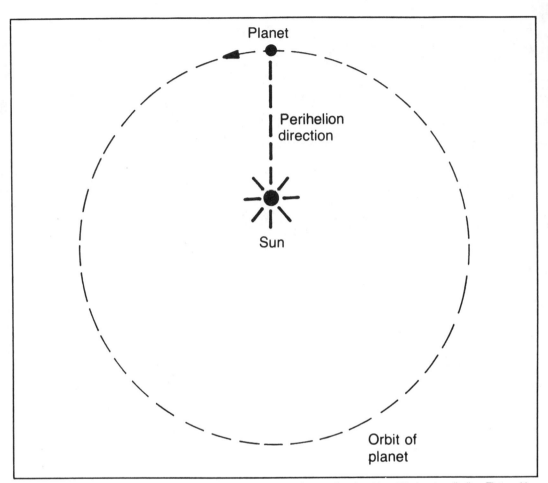

Fig. 7-26. When a planet passes the point in its orbit that is closest to the sun, the planet is said to be at perihelion. The position of this perihelion changes slightly with each orbit, according to the general theory of relativity, but Newton's physics cannot provide any reason for this change.

ity to the sun maximizes the rotation of the perihelion, should theoretically display a change of 43 seconds of arc per century. This amount can be measured. For the planet Venus, the predicted change in the perihelion is 8.6 seconds of arc per century. For our own planet, the theoretical value is 3.8 seconds of arc per century. These values are also within our capability to measure.

According to observation, scientists have obtained a very close agreement with the values predicted by the general theory of relativity for Mer-

cury and Earth; for Venus, the observations have not yielded a reliable value, and we have no means for comparison.

The reason for the movement of the perihelion of a planet lies in the fact that the sun rotates on its axis once every 21 days, with respect to the distant totality of masses in the cosmos.

RED SHIFT CAUSED BY GRAVITATION

We have already briefly discussed Doppler shift, the change in wavelength of a light beam or radio

148

signal caused by radial motion (Chapter 2). But this sort of frequency change can occur also for another reason. Radiant energy changes in intensity in the proximity of a gravitational field, and this change persists after the energy has left the field. The wavelength will get longer as a beam of light leaves a gravitational field, or when it travels against an acceleration force. The wavelength will get shorter as a beam of light enters a gravitational field, or when it travels along with an acceleration force. The former is referred to as a red shift, and the latter as a blue shift.

Distant stars have a characteristic spectral pattern when their light is passed through a prism. Various elements cause absorption of certain wavelengths in the spectrum, and the absence of these discrete wavelengths shows up as dark lines in the spectral image (Fig. 7-27A). We are thus able to easily see if a red or blue shift has occurred; an increase in energy will show up as a movement in the positions of the absorption lines toward the blue end of the spectrum (Fig. 7-27B), and a decrease in energy will cause a shift toward the red (Fig. 7-27C).

Red and blue shifts may, of course, be the result of radial motion of a celestial object. But, as the light of the star leaves that star, a red shift occurs in addition to any Doppler effect that may be present, since the light must move against a gravitational field; a blue shift occurs as the light from the star falls in to the earth, along with the gravity of our planet. The blue shift resulting from our own earth's gravitation, f_{earth}, is much less pronounced than the red shift caused by the star's gravitation,

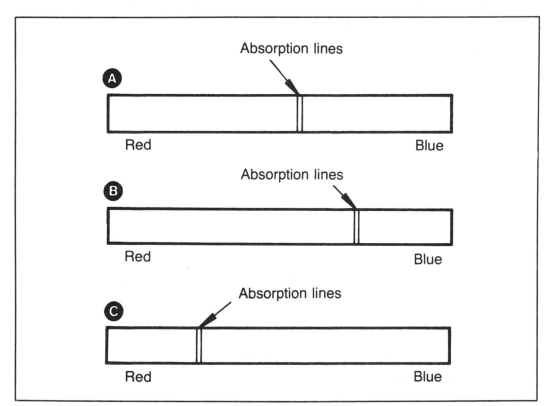

Fig. 7-27. Red and blue shift of the absorption lines in the spectrum of a distant object. At A, no shift; at B, blue shift; at C, red shift.

f_{star}. Thus, as long as the star has zero radial velocity, the net spectral shift will be toward the red (Fig. 7-28).

The practical difficulty in experimentally verifying this red shift is twofold. First, how are we to determine whether a red shift, observed in the spectral display of a distant star, is caused by radial motion or gravitation? A retreating star would produce a red shift indistinguishable from the shift caused by $f_{star} - f_{earth}$. Second, convection currents in the surface gases of a star might create a blue shift sufficient to cancel, at least partially, the red shift caused by $f_{star} - f_{earth}$. Still, experiments have been attempted to observe this gravitational red shift from distant stars. Rough quantitative results have been obtained, by observing a great number of stars whose average radial velocity should be zero, and by estimating the effect of convection currents. Although these data are somewhat imprecise, they are in fair agreement with the red-shift values predicted by the general theory of relativity.

The red shift caused by the gravitation of our own sun has been observed by measuring the value near the rim of the disk, where convection currents in the solar gases occur only in directions sideways to us (Fig. 7-29). The results are in good agreement with Einstein's predictions when the red shift is observed near the rim of the disk; but the blue shift caused by convection reduces the amount of red shift observed as observations are made closer and closer to the center of the sun's disk.

The displacement of spectral lines through the gravitational field of the earth has also been measured using techniques of extreme precision, over distances of only a few meters. The results are in close agreement with the values predicted by the general theory of relativity.

GRAVITY WAVES

Newton's theory asserts that the effects of gravitation move instantaneously through space. Einstein found this idea hard to accept, since one of his fundamental principles requires that the speed of light be an absolute limiting speed, not only for radiant energy, but for all effects: light, time, magnetism, ideas, and gravitation. As a matter of fact,

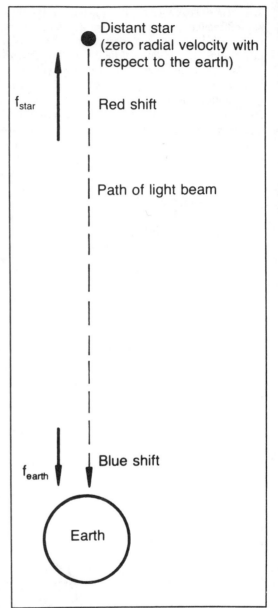

Fig. 7-28. According to the general theory of relativity, the light from a distant star having zero radial velocity should be red-shifted. As the light leaves the star against the gravitation f_{star}, it is red-shifted. As the light falls to the earth through the gravitation f_{earth}, it is blue-shifted, but not as much as it is red-shifted by the field f_{star}.

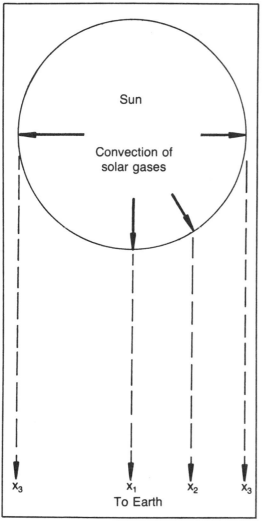

Fig. 7-29. To get a true picture of the gravitational red shift from the sun, we must observe light from the edges of the disk (points x_3). If we look near the center, at point x_1, or at an intermediate point, x_2, the convection of solar gases will cause some blue shift that will partly cancel the observed red shift.

"thought transfer" experiments were actually attempted, in order to perhaps verify that even extrasensory effects could move only with the speed of light! (Such experiments could not be done, however, on a reliable enough basis.)

If gravitational effects do move at a finite speed, then perhaps there exist gravity waves, just as it is possible to have electromagnetic waves. Distant events in the universe might cause a kind of "gravitational shock disturbance" that would manifest itself as gravity waves. These waves were sought by experiment, and they were indeed found.

Gravity waves occur as distortions in space—changes in the curvature of the continuum—in an oscillating pattern. This effect is illustrated in Fig. 7-30. Gravity waves are equivalent to a constantly changing acceleration force. At certain times, a wave "crest" or "trough" will occur at the measuring point (A), producing one kind of distortion; at other times the passing wavefront is at an intermediate phase (B), producing a different distortion. These waves pass at a velocity equivalent to the speed of light, c. They have a definite wavelength, λ, and a measurable frequency, f, such that $c = f\lambda$. This is the same wavelength-to-frequency relation that applies to electromagnetic fields.

The apparatus used to detect gravity waves consists of a large, massive cylinder of metal, with very precise dimensions. Changes in the force on the cylinder, and in its spatial length, are detected by means of sensitive instruments. So sensitive are these instruments that even the gravitational change caused by a person walking by will register significantly! Thus, to be sure the gravitational disturbances are coming from outer space, two cylinders must be used, located far enough apart so that any random disturbance near one cylinder will not occur near the other. Then any effect that occurs simultaneously, to an equal extent at both locations, can be assumed to have an extraterrestrial origin.

According to Newton's theory of gravitation, gravity waves cannot take place, because the speed of gravitation is infinite, and no meaningful frequency-to-wavelength relation can exist. For wave action to be possible, the speed must be finite. But gravitational waves have been observed, coming from certain remote celestial objects. So the speed of gravity must be finite. These waves may be thought of as "ripples" in the continuum of space, exactly like the waves on a pond that occur around

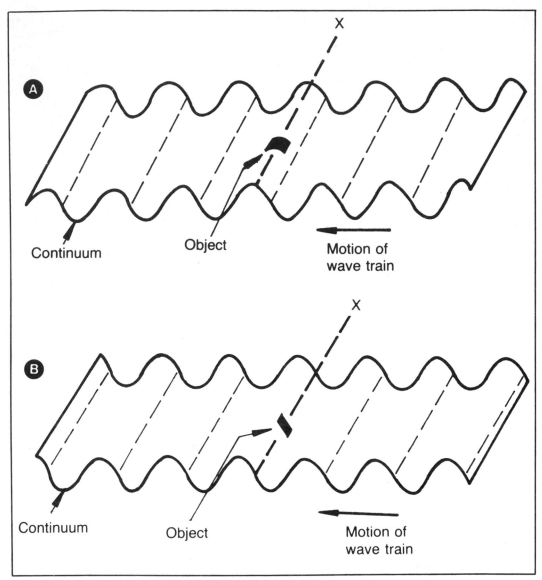

Fig. 7-30. Gravity waves passing an object. At A, the object is at the "crest" or "trough" of a wave. Its position line is given by X. At B, a short moment later, the object is past the wave peak. Passing gravity waves cause changing forces on the object, and changes in its spatial proportions. These changes can be measured by experiment.

the impact point of a stone thrown into the water, but in one more dimension. The origin of gravity waves in the far reaches of space will be discussed in the next chapter.

THE UNIFIED FIELD THEORY

Gravitation and electromagnetic effects show many similarities. We have just seen one such analog, namely, wave action; but there are others. Opposite electric charges attract and like charges repel. The same is true of magnetic poles. There are electric and magnetic fields, just as there are gravitational fields. We have seen that a gravitational field is a curvature in the geometric structure of space. But electric and magnetic fields can be mathematically described in much the same way as gravitational fields. Is one kind of field really of different origin than the others, or can all fields perhaps be explained according to the same common base? Must we invent a different spatial continuum to explain each different kind of force field? This question bothered Einstein. It is a source of puzzlement among scientists to this day.

Einstein sought a uniform explanation, based on fundamental concepts and not patchwork contrivances, for the effects of gravitation, and all force phenomena in general. He hoped to give a precise explanation for the existence of electrons and protons in the atom, their charge, and in fact a model for all of quantum mechanics, based on one set of "world laws." This elusive goal became known as the unified field theory. Work is still being done today toward this objective.

IS SPACE REALLY CURVED?

We have spoken here of rather bizarre effects caused by gravitation. According to the theory of general relativity, the geometry of space becomes non-Euclidean, that is, curved, both locally and over all the universe. Is this curvature real? Or is it just fancy mathematical trickery? There are those who will assert the latter, simply refusing to accept the plausibility of something they cannot readily envision. But, there is already good evidence (aside from the effects we have discussed in this chapter) for the theory that space actually is curved. Astronomers believe they have found evidence of "black holes" in our own galaxy and in distant galaxies. These strange objects are surrounded by gravitation so intense that space folds all the way around on itself, creating a tiny, separate universe attached to the larger continuum. This gives rise to the idea of a one-way membrane—a point across which it is possible to get in, but not out. The general theory of relativity also predicts a finite but unbounded universe, like the surface of the earth. Modern cosmologists accept that this model of the geometry of the universe may well be correct.

To those who maintain that the curvature of space must be pure mathematical fiction, we might reply that we know only what we are able to observe, and all our observations to date have either confirmed the general theory of relativity, or else they have been too imprecise for conclusions to be drawn. Some people still believe the earth is flat. We might ride an airplane or ocean vessel all the way around the world with such persons, showing them how we arrive back at the starting point and meet the same people, all of whom remember us, and challenge the "flat-earthers" to explain *that*. But, they might argue that there must be exact duplicates of all the things on earth, extending infinitely in all directions, and repeating themselves over and over without end. We might then retort, in elegantly Einsteinian fashion, that their world and ours is really the same. Both models give the same results, and thus they must be identical!

The question of whether space is really curved thus becomes trivial. It behaves exactly as though it is curved, and so we might as well take the simplest approach and consider that it is.

The theory of curved space leads to some extremely strange possibilities: anomalies in space such as black holes, one-way membranes, intersecting universes, and time standing still. Let us now explore these things. Beyond the imaginations of even the majority of science-fiction enthusiasts, cosmologists of today are proposing, as reality, such phenomena.

Chapter 8

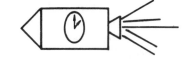

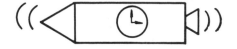

Anomalies in Space

T HE IDEA THAT SPACE BECOMES NON-EUCLIDEAN in the presence of a gravitational field, or under conditions of extreme acceleration, is of little practical consequence under most conditions. The acceleration attainable by man-made space vessels cannot be great enough to bend the geometry of space very much without crushing the passengers to a nearly homogeneous mass of atoms. The same holds true in a gravitational field: A source of gravity strong enough to cause significant curvature of space would also kill anyone who dared to venture close enough to find the curvature. In the gravitational field near the sun, our bodies would be put under stress they could not withstand; yet the sun causes only a tiny amount of curvature in the continuum. While the general theory of relativity explains the behavior of the universe in a more accurate way than the classical models, the difference is minor until we are given conditions of sufficiently strong gravitation or acceleration. The rotating earth situation that we discussed in the previous chapter is not quite the sort of arrangement we need

to illustrate the true significance of non-Euclidean space. A far more interesting situation occurs when a star dies. If a star is large enough, the gravitational pull increases at its surface to the point where extreme distortion of time and space takes place. The homeomorphism between the universe near the star and the space outside the star may disintegrate, resulting in a "hole."

THE DEATH OF A STAR

Many millions or billions of years from now, our own sun will begin to run out of the hydrogen fuel from which it has created so much radiant energy by means of nuclear fusion. We think of our sun as just about the most absolute thing in our lives. There may be earth tremors, volcanoes, and tidal waves; or we may interfere with our own environment, such as by destroying the ozone layer in the stratosphere or incinerating our continents with miniature suns. But whatever the terrestrial changes that lie in store for us, we also have the inner confidence that we can survive. The death of the sun would shake that confidence.

As the sun's supply of hydrogen fuel becomes exhausted, the internal pressure within the sun will decrease because of lowering temperatures. There are several variations of the theory concerning events to follow. But, it appears likely that the sun will shrink in size until the internal pressure increases once again, causing helium (the by-product of hydrogen fusion) to undergo further nuclear reactions and form still heavier elements. But, eventually the helium supply will also run out, and the elements that remain will eventually become too heavy to support fusion reactions.

Whatever the exact details of this process, there is one overriding fact: The energy supply of the sun is not unlimited. There will come a time when the sun will no longer shine. There will still be matter left in our star at this time; not all the matter can be turned into energy, because elements can get no heavier than a certain maximum. Surely, if every last proton, neutron and electron were used up, our theory would be simpler. But, when the star we call "Sun" is dead, there will be a small sphere remaining, about the size of the earth, but many times heavier than our planet. The material left in the sun after it has died will be so dense, in fact, that a teaspoonful would weigh tons on earth.

The gravitational field near the surface of this tiny, massive ball of heavy elements will be many times greater than even the field presently at the surface of our sun. If we were able to survive, somehow, until the death of our sun, and were then to carry out the experiment of Arthur Eddington and his colleagues at the beginning of the twentieth century, we would see much greater relativistic distortion of the background of stars. The rays from distant stars, passing near the dead sun, would be bent much more than they now are. Peering through a telescope at the little black disk we would then regard as the "sun," we might even be able to observe, as the sun passed across the background of other objects in space, a pronounced oddity. It would look as if the sun were embedded in some sort of strange refracting lens (Fig. 8-1).

This would be the end of our sun. More massive stars would die to form more massive and more dense "black dwarfs," with greater gravitational pull and hence even more distorting effects on the surrounding universe. If a star were sufficiently massive, the force of gravitation would grow to incredible proportions as the matter collapsed. The atoms, comprised mainly of empty space in their familiar state, would be crushed. Electrons would be driven into the nuclei, combining with protons to form neutrons. The nuclei themselves would continue to be crushed until the star consisted of just a huge conglomeration of neutrons; very little space would remain in between them. Like a great mass of warm chocolates, these neutrons would merge together, to form what might best be described as one gigantic neutron—the exact replica of its former constituents, but on a larger scale.

At the surface of an object like this, commonly called a *neutron star,* a man would be killed instantly, spread into a paste of atomic particles and further modified thereafter into a neutron ooze, like the star itself. Any piece of matter meeting the ill fortune of falling into the neutron star would end up the same way. The gravitation would be tremendous, and if the neutron star became large enough, the geometric distortion of space would reach such magnitude that a peculiar thing would happen: The escape velocity from the star would become greater than the speed of light (Fig. 8-2).

GRAVITATIONAL COLLAPSE

A star of great size awaits a more violent future than our sun. It is the large star that sometimes flares into a supernova, temporarily attaining thousands or millions of times its previous brilliance in the process of exploding. (The Crab Nebula is the cloud of matter created by an exploding supernova in the year 1054 A.D.) The large star also awaits a more bizarre final state. It may shrink down to a neutron star so dense, and with such powerful gravitation, that photons emitted from the surface cannot escape into space, but instead will fall back to the star. When this occurs, we say that the neutron star has become a "black hole." It indeed looks black; we cannot see any radiant energy coming out of it. Any light that comes too close will be pulled down to the

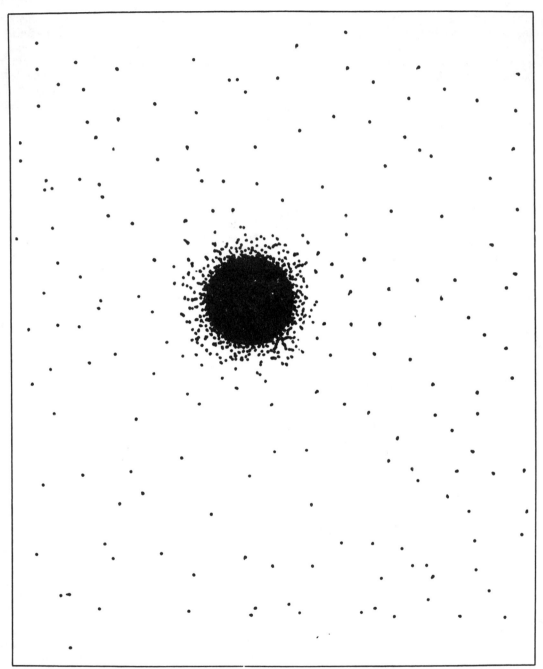

Fig. 8-1. Distortion of the background of stars near the sun in its final, highly dense state. The curvature of space in the sun's vicinity will be much greater than it is at the present time.

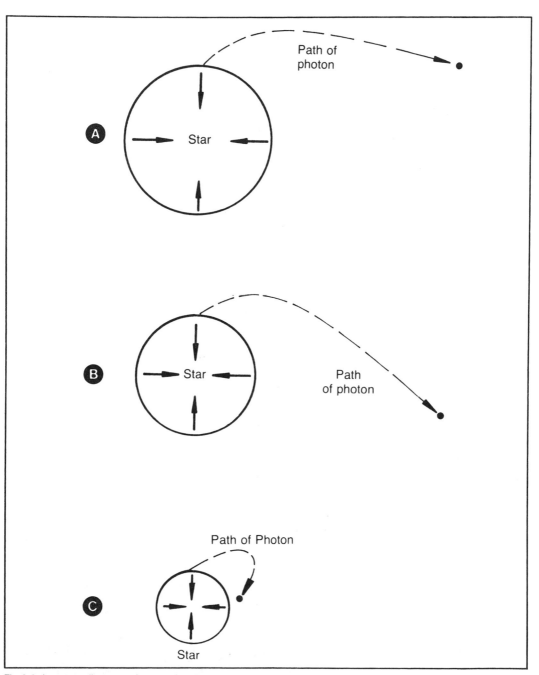

Fig. 8-2. As a star collapses, a photon emitted by the star encounters greater and greater difficulty in escaping. At A, the path of the photon is curved greatly. At B, the photon just barely escapes. At C, the photon falls back to the star.

star, never to return to the outside universe again.

As a neutron star shrinks to smaller and smaller size, it will seem to disappear at a certain point. Its radius at this instant—at the exact moment the escape velocity becomes greater than the speed of light—is called the gravitational radius or Schwarschild radius (the latter term is named after the physicist who first put forth the idea of a black hole using Einstein's general theory of relativity).

For a neutron star at its gravitational radius, beams of light sent directly upward will almost escape. But, just as a rocket ship hurled away from earth at a speed just short of the escape velocity will fall into orbit, so the photon will fail to reach the outside universe. If the photon is not emitted at a sufficiently large elevation, it will fall back to the star (Fig. 8-3). Since nothing can travel faster than a photon, we may then consider that a neutron star, once having shrunk to within its gravitational radius, is sealed off from the rest of the universe. It is out of the question to consider leaving; to do so would require an infinite amount of energy.

The curved paths of photons (however distorted they are as shown in Fig. 8-3) near a black hole are nevertheless geodetic paths through space. From the viewpoint of someone standing on the surface of the neutron star (assuming he could survive the gravitational pull), the photons would appear to travel in straight lines. Space would thus appear greatly distorted.

Imagine that you were able to stand on the surface of a collapsing neutron star, and witness its retreat into gravatational oblivion. What would you see?

At first, everything would appear to be quite normal. The surrounding sky would contain the usual plethora of stars. You might even recognize some of the familiar constellations if the collapsar were near enough to our own sun. But as the intensity of the gravitational field became greater and greater, and the geometric distortion of space increased, the sky would change. New stars, previously invisible because they were below the horizon, would appear to rise upward from the horizon at all points of the compass. All the stars in the sky would seem to be moving upward toward the

zenith; the stars closest to the horizon would appear to move the most, and the stars at the zenith would remain fixed (Fig. 8-4A).

As the gravitational collapse continued, the rate of change in the surroundings would increase.

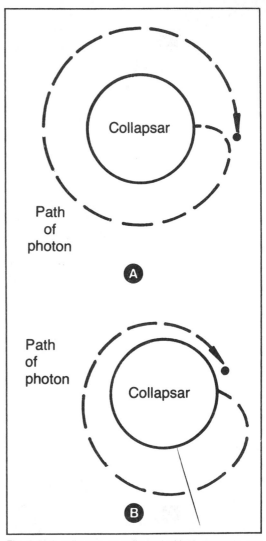

Fig. 8-3. A collapsar, or collapsed object, prevents any photons from ever escaping. The best a photon can achieve is an orbit around the collapsar (A); but if the photon is emitted at an angle other than directly upward, it will eventually return to the surface of the collapsar (B).

Finally the horizon would heave upward, and you would get the feeling of being at the bottom of a huge bowl (Fig. 8-4B). Now things would change with extreme speed; the end would be very near. The whole horizon would retreat to the zenith, and close off your view of the outside heavens completely. At that instant (Fig. 8-4C), you would be beyond the point of no return; you would never be able to escape. The star would be inside the gravitational radius, and it would be severed from the rest of the world. Looking directly upward, you might even see yourself for a tiny instant, perhaps just long enough for you to note the look of sheer astonishment on your face. Then the whole surface of the neutron star would come crashing down on you from all directions. The new, autonomous universe created by the gravitational collapse would contain only neutron ooze, and no surface.

AN OUTSIDE VIEW

Now, imagine this same situation as seen by someone far enough away from the star so as not to be sucked into its grasp. At first the star would seem to rapidly shrink. But, as the gravitation became extreme in the immediate vicinity of the collapsar, time would go more and more slowly. A clock situated at the surface would seem, as seen from the outside, to slow down to a complete stop as the star reached the gravitational radius.

Finally the collapsar would disappear from view. We might see the poor soul at the surface staring helplessly upward, his expression frozen by the infinite time-distortion factor that prevails at the entrance to a black hole.

This time distortion, caused by the gravitational field of a collapsing star, results in its never completely disappearing from the outside universe. Visually, it will appear to wink out because the time distortion becomes so great that individual photons arrive at large intervals, rendering the object too dim to be seen. But the gravitational effect of a black hole persists forever in the outside world. As our clocks continue onward, the clock on the surface of the collapsar appears to come to a stop as the gravitational radius is approached. This effect is shown in Fig. 8-5. The observer on the surface of the collapsing star would see time go on quite normally, but in the brief moments prior to the severing of his universe from the outside world, he would see all of eternity in that world pass by.

The collapsing star thus creates a non-homeomorphism between two universes. Time-wise, a single instant in the collapsar universe corresonds to the entire hereafter in the external world; spacewise, the surface of the star folds down to one single point. Both of these relations are not one-to-one correspondences, but rather "infinity-to-one." A collapsar thus "tears" space apart.

GRAVITATIONAL RADIUS VERSUS MASS

The more massive a star, the greater its chances of becoming a black hole. Larger and larger stars will form larger and larger black holes.

Any object can, in theory, be sufficiently squeezed down so that it gets smaller than the gravitational radius, and creates a black hole. Even the earth can be sufficiently compressed, although the task might prove challenging (it would have to be crushed to the size of a grape). The sun would have to be compressed to a diameter of several kilometers; as the amount of matter in an object increases, the gravitational radius increases in direct proportion. If all the matter in our Milky Way galaxy were moved sufficiently close together, the resulting black hole would be very large indeed. The volume of an object increases much more rapidly than the radius, however, and consequently large black holes need not be nearly as dense as small ones.

The black hole resulting from the collapse of a great many stars together, or from the collapse of an entire galaxy, would not be nearly as concentrated as the neutron-star black hole that is the final product of a single star's demise. Inside a massive black hole, in fact, human existence might actually be carried on. If a black hole is large enough, it would be possible to venture inside its gravitational radius—also known as the "event horizon" because of the infinite time-distortion that takes place there—without being crushed to death by gravitation or neutron ooze. We might then explore the

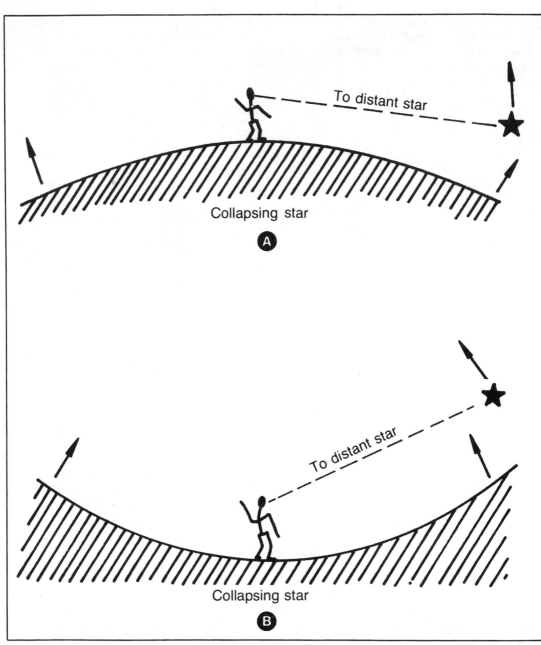

Fig. 8-4. Standing on the surface of a collapsing object, space seems to close in on an observer. At A, a distant star is seen, but the horizon seems to be lengthening and the star rising. At B, the surface of the collapsing object looks as though it is curved upward, and the star is much higher in the sky. At C, the observer is trapped and the collapsar is about to crush him out of existence.

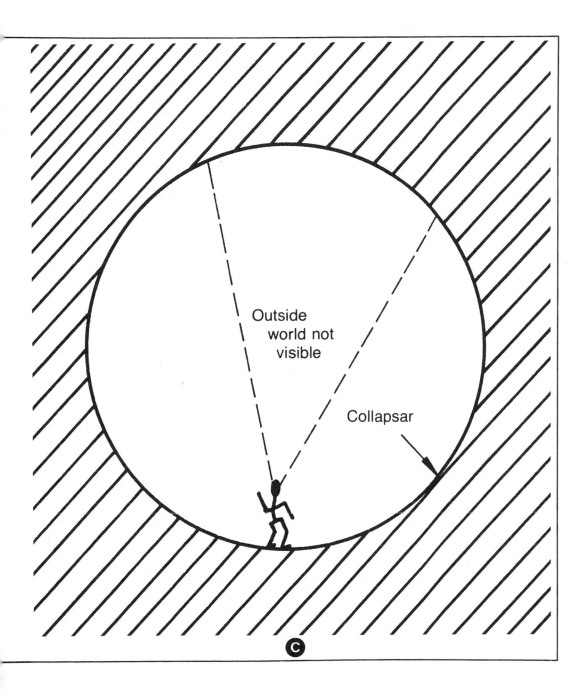

Outside world not visible

Collapsar

©

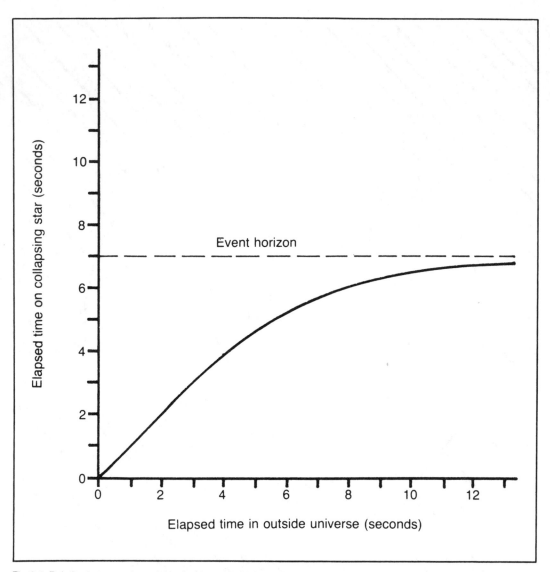

Fig. 8-5. Relation between elapsed time inside and outside a collapsing section of space. From the outside, the event horizon or gravitational radius seems never to be quite reached. From the inside, all of eternity for the external world seems to pass in an instant as the object shrinks within the event horizon.

interior of the black hole, beyond eternity! But there would be no hope of returning to the old universe.

Inside an extremely large black hole, individual stars might still retain their identity. If such black holes exist, having many times the mass of our galaxy, might it be possible that there are planets orbiting some of the stars? Could it be that there is intelligent life on one or more of those planets? Might those beings be pondering the nature of their little universe in the same way we wonder about ours?

HOW COMMON ARE BLACK HOLES?

Black holes are interesting in theory, but it is difficult to imagine such things as the pronounced geometric distortion of space and total stoppage of time that is supposed to occur inside them. Are there really such things? Do stars really get that dense as they die, or is there perhaps some as-yet unknown factor that prohibits such bizarre events from actually happening? There is some evidence that black holes actually exist in the far reaches of space. We certainly have never seen one here on the surface of the earth, and hopefully we never will! Finding good evidence of the existence of black holes, however, is good support for the general theory of relativity. It also makes the universe far more interesting.

The motions of stars in space are governed by the effects of their mutual gravitation. If there are no black holes, then the motions of the stars ought to be predictable based on the locations and sizes of nearby visible stars. In most galaxies, but especially in elliptical and spherical galaxies, there appears to be too much motion among the visible stars. Such galaxies seem to harbor disproportionately large mass, based on the radiant energy coming from them. If all, or almost all, the mass in a galaxy can be accounted for as visible stars, then we would expect the luminosity-to-mass ratio to be about the same as for ordinary stars such as our own sun. But, in fact, the ratio is only about one-sixtieth that of the sun. While we might dismiss a small discrepancy, a sixty-fold difference seems too great. Where is the extra mass in such galaxies?

Perhaps there is a much greater concentration of stars at the centers of these galaxies than we would be led to believe from visual examination, or by observation with radio telescopes. It may be that the extra energy is somehow confined within the centers of these galaxies. But how could that be? Why would the energy not eventually escape into space? One plausible answer is that the central regions of these galaxies have become so highly compressed that they are black holes. This would not require a density all that great; if billions of stars were brought together within a sufficiently small volume of space, the resulting black hole would have about the same density as the constituent stars. We would not see any of the energy radiated by such a galactic core; it would all be held within the gravitational radius. But the black hole at the center of a galaxy would act quite effectively to keep the rest of the stars in the galaxy together.

Clusters of galaxies, held together by their mutual gravity, also display masses that appear too large for the amount of radiant energy they put out. There are apparently invisible galaxies among the visible ones, and some estimates actually imply that the invisible galaxies outnumber the ones we can see! Is this observation simply the result of imprecision in our estimates of luminosity and mass? Perhaps, but not necessarily. It could be that there are entire galaxies that have been drawn within their gravitational event horizons, leaving only their gravity as evidence that they still exist.

This evidence for the prevalence of black holes in the universe, while tantalizing, is not very conclusive. There are too many variables. It is possible that there is more stray matter between the stars and galaxies of space than we think; this would reduce the luminosity-to-mass estimate for distant objects in the universe. Although it is hard to dispense with a sixty-fold difference in this way, we must still keep in mind that such an error might occur; we are working with a fantastically complicated creation. We need still more data to imply with surety that black holes actually exist.

GRAVITATIONAL SHOCK WAVES

We briefly discussed gravitational waves in the preceding chapter. Such waves consist of ripples in the continuum of three-dimensional space, very much like the waves that occur on the surface of a smooth body of water in ever-growing circles around an impact disturbance. What might cause such gravitational waves?

The effects of a gravitational field, insofar as the geometric distortion of space is concerned, extend for indefinite distances in all directions around a celestial mass. Of course, the greater the distance

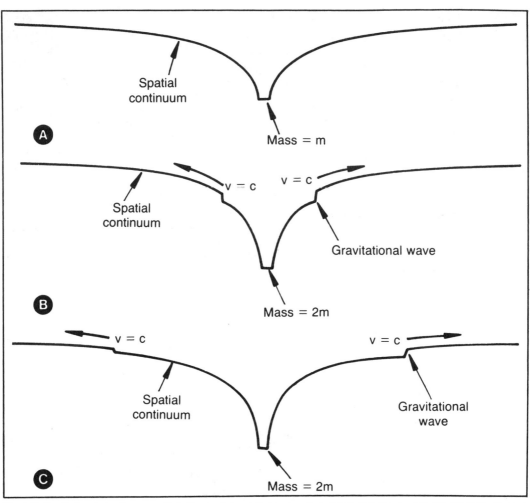

Fig. 8-6. Gravitational shock waves. At A, the spatial curvature around an object of mass m; at B, the gravitational wave produced by a sudden doubling of the mass; at C, the shock wave at a later moment. The velocity of the wave, v, would be equal to the speed of light. Instruments to detect gravitational waves have been built, and the waves have been observed coming from the center of our galaxy.

from the mass, the smaller the effect of the gravitational field. But, no matter how far away we get, there is always a little bit of spatial curvature. Even a tiny speck of dust creates a space "warp" with indefinite extent!

What if the mass of an object were to suddenly change? The intensity of a gravitational field varies directly with the mass of the object responsible.

Therefore, a change in the mass of an object would result in a change in the degree of spatial distortion surrounding that object. This change would only be capable, however, of being propagated at the speed of light. Consequently there would be a disturbance in space in the form of an ever-growing sphere, where the curvature would suddenly become different. Figure 8-6 is a dimensionally reduced illustration of this effect. The "pulse" that travels outward is a gravitational wave.

How might the mass of an object suddenly change? It is possible that two stars might collide, resulting in a sudden change in the mass of either star, and this event might cause repercussions in the form of gravitational waves. Such chance encounters are highly improbable, since there is far more empty space in the universe than space occupied by stars. We would not expect to find stellar collisions very often.

Might a star suddenly disappear during a gravitational collapse? Not as viewed from the outside universe; the amount of time distortion becomes infinite as any object collapses to the gravitational radius, and this ensures that the black hole will persist forever. (We do not know exactly what happens, though, as seen from *inside* the black hole.) But, a black hole might suck other stars into it, thereby growing in size in discrete spurts.

Suppose there is a large black hole somewhere in the midst of many stars, such as at the center of a galaxy. (We might expect to find a black hole in a concentrated region of stars much more often in an area where there are not many stars, since the greater material density would increase the chances of a mutual gravitational collapse.) Such a black hole would have a tendency to attract nearby stars, on account of its intense gravitational field, eventually capturing them one at a time. This kind of collision would be much more likely, supposing the galactic center really harbors a black hole, than a chance meeting of two stars. Each time a star was pulled down into the singularity, the mass of the black hole would increase, and we might expect to detect a burst of gravitational radiation.

The apparatus needed to detect gravity waves has already been briefly discussed (Chapter 7); it consists of a large cylinder of metal with very sensitive devices attached, for the purpose of measuring tiny changes in the physical proportions of the cylinder. Beginning in 1969, experiments were conducted in an attempt to discover gravitational radiation. The scientist who supervised these activities was Joseph Weber of the University of Maryland.

Results of Weber's experiments proved very rewarding. Many disturbances were observed which could not be attributed to earthly effects. The source of most of the gravitational waves seemed to be in the general direction of the center of our Milky Way! Something cataclysmic was evidently happening there. It may well be that there is a great black hole there, and that the gravitational radiation is the result of periodic increases in its mass as individual stars fall into the singularity, disappearing from the outside universe forever.

STRANGE BINARY STARS

Certain objects in space provide more readily visible evidence for the existence of black holes. Binary stars are quite common. Occasionally, the plane of the stars' mutual orbit lies directly in line with our solar system, and the stars alternately eclipse each other. We then see a periodic variation in the brightness of the binary star system. These variable stars have, in fact, been known for some time. Generally, one star in a binary system is considerably more luminous than the other, and therefore the fluctuations in brightness may be quite large (Fig. 8-7).

Occasionally we observe a binary star system in which the eclipse of one star results in almost total cessation of light emission from the system. This might well be caused by a binary star system consisting of one luminous star and one collapsed star (Fig. 8-8). A sufficiently massive black hole could "suck in" much of the electromagnetic energy passing near it.

By observing the characteristics of binary star systems with one invisible member, we can obtain rough estimates of the masses of the two stars. Quite often, strange conclusions are reached, implying that the supposedly smaller, invisible star has much greater mass than the bright star. An ordinary black dwarf star would not cause significant changes in the luminosity of the system as it passed in front of a large, bright companion. Such an eclipse would be comparable to the changes in the brightness of our sun when it is eclipsed by the planet Venus. The effects of a black hole, however, would be much greater, because light would literally be pulled in toward the collapsar. Such binary systems, whose characteristics are nicely explained by this model, have been observed in our own galaxy.

Some binary systems have been observed in which one invisible companion seems to be tearing away pieces of the larger, visible member. A tiny dwarf star would hardly be expected to do that, unless it had extremely great mass. Observations of Cygnus X-1, an X-ray star in the constellation Cygnus, have led scientists to believe that the invisible member has ten times the mass of our sun, but has very small diameter. Such an object is an excellent candidate for a black hole.

QUASI-STELLAR RADIO SOURCES

Just recently, astronomers noticed strange objects at the farthest reaches of the observable universe. As telescopes were built to probe more and more

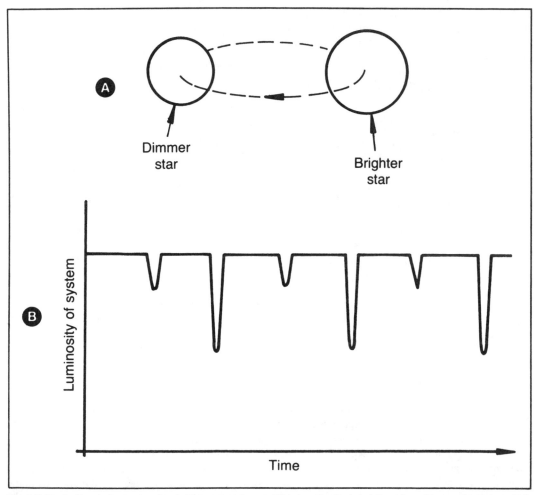

Fig. 8-7. Fluctuations in the luminosity of a binary star whose orbital plane lies in almost direct line with our solar system. At A, the star system is shown; at B, the luminosity is graphed as a function of time. First the dimmer star is eclipsed (small drop in total luminosity), and then the brighter star (larger drop in luminosity).

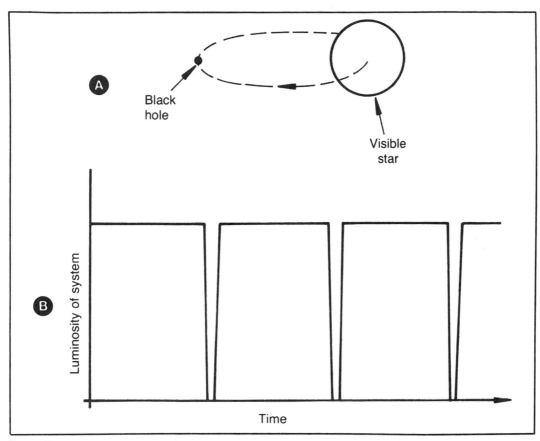

Fig. 8-8. Fluctuations in the luminosity of a binary star with a black hole. At A, the star system is illustrated, showing the plane of mutual orbit between the visible star and the collapsed star. At B, the luminosity is graphed as a function of time. No change is observed when the black hole is eclipsed by the visible star, but the brightness drops almost to zero when the visible star passes behind the black hole.

deeply into the cosmos, more galaxies and clusters of galaxies were found. But eventually, when radio astronomy became sophisticated, certain star-like objects were found to be as distant as the most remote galaxies. Through the most powerful visual telescopes, these objects appear as points of light just like ordinary stars in our own galaxy. In fact, they were mistaken for nearby stars until radio telescopes were turned on them. Then it became evident that they were not like other stars. They put out far too much energy.

These quasi-stellar radio sources, also called quasars, display fairly rapid changes in the intensity of their radiated energy. Thus we know they cannot be galaxies, since galaxies contain billions of stars and the chances of all of them varying their brightness together are ridiculously small. Yet, estimates of the distances to quasars places them very far away indeed. They cannot be ordinary stars either. What are they?

Apparently the quasars are very compact objects. One model proposes that the quasars are galaxies nearing a state of total gravitational collapse. A large, rapidly rotating black hole would be capable of putting out great amounts of radiant energy in its plane of rotation, as the last stars are pulled into the gravitational event horizon. A process such as this could go on for many millions of years. Some support of this model arises in considering the fact that, because of the great distances to the quasars, they must represent events in our universe that took place billions of years ago. The light and radiant energy that we see from them departed in the distant past. We should then expect that there are many "black galaxies" nearer us, whose collapse has been completed. As we have seen, there is strong evidence for such invisible galaxies in many of the observable clusters.

Might our own Milky Way galaxy, if it harbors a black hole at its center, face an ultimate demise into total gravitational oblivion? It is quite possible. All galaxies may eventually end up as gravitationally collapsed matter. This might, in fact, be the fate of all matter in the universe. At present, we can only speculate about this, but we can at least remain confident that the universe will not meet this end in our lifetimes. As the science of cosmology, born only in this century as a scientific rather than purely philosophical pursuit, grows more advanced, we may gain better understanding of where our universe is headed.

COLLAPSARS AND SPACE TRAVEL

Astronauts have traveled to the moon, and unmanned probes have been sent throughout much of the solar system, and not yet have any of our prefabricated celestial objects been unfortunate enough to run into a black hole. Still, we do not know how common black holes actually are. A tiny black hole, such as one earth mass, is not a particularly likely product of a gravitational catastrophe; only large stars are expected to evolve into black holes via that process. Until men are capable of fairly rapid interstellar excursions, it is not likely that any spacecraft will encounter a singularity.

If there were a black hole within our own solar system—for example between the orbits of Saturn and Uranus—we would expect to observe its gravitational effects on those two planets even if the black hole were only a few millimeters across. A black hole just one centimeter in diameter would mass the same as the earth, and though we could not

hope to see it with our telescopes, it would give its existence away by causing perturbations in the orbits of the nearby planets. The larger the black hole, the greater this disturbance would be. A collapsed star would not allow our solar system to exist as it does; the orbits of the planets would be greatly distorted from the nearly perfect circles they are. We may thus venture with confidence into all of our solar system. No strange planetary behavior, suggesting the existence of a black hole, has been seen.

Imagine that we are on a journey among the stars, and accidentally happen to come close to a black hole. We might notice an odd arrangement of the background stars, caused by the gravitational effect on the geometry of space in the vicinity of the singularity; with sensitive instruments for measuring tidal forces, we might sense the proximity of a large celestial mass. But, if there were not many stars visible, and the black hole was for some reason very small, there is some danger that it might pass unnoticed until we came very close to it. Then our ship would suddenly change its course and speed. If we were unlucky enough to be headed too close to the event horizon, we could be drawn into the black hole.

A fateful scenario such as this is very improbable, even if there are a lot of black holes out there. First, the sheer statistical likelihood of getting close enough to a collapsar to be greatly affected is essentially nil. Space is vast indeed; the largest stars occupy only a minute proportion of its volume. Even if black holes outnumber visible stars by five to one, it would be a fantastic coincidence if a space vessel were ever to run into a singularity. Further, we have every reason to believe that sufficiently sensitive gravitational detection instruments can be built to warn us of any change in the geometry of space. (We would need these devices for navigation, anyway.)

It is possible that neutron stars or black holes might provide a means of attaining greater speed on an interstellar journey. Interplanetary probes have utilized the gravitation of one planet to gain additional speed on the way to the next; similar maneuvers are possible in interstellar travel. A black

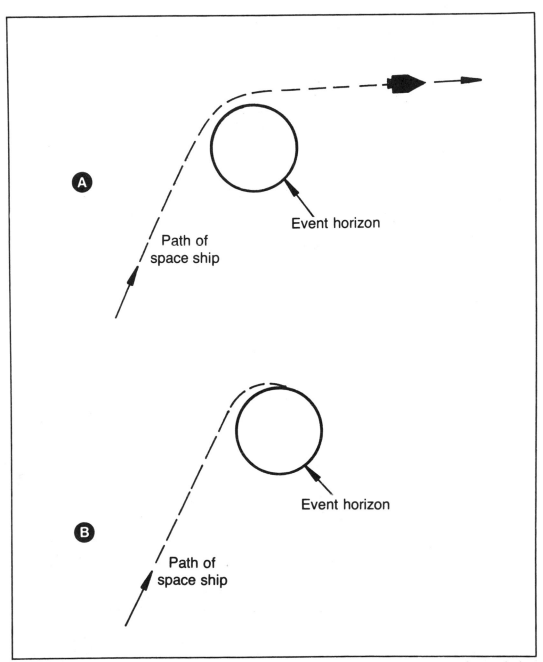

A

Event horizon

Path of
space ship

B

Event horizon

Path of
space ship

Fig. 8-9. We might someday use a black hole to give us a speed boost in interstellar travel. At A, passing near the event horizon changes the direction and speed of our vessel. But we had better not get too close or we will be pulled in (B).

hole would obviously provide a much cooler near-pass environment than a normal star. The gravitational acceleration would also be considerably greater. In the extreme field near the event horizon of a black hole, however, the margin for error would be small (Fig. 8-9). A tiny miscalculation in the speed of the ship, or in its direction, could result in destruction.

COLLAPSARS AND TIME TRAVEL

The intense gravitational field in the vicinity of a black hole causes appreciable time distortion. At the event horizon itself, as we have seen, the time-distortion factor becomes infinitely large. Just outside the gravitational radius, the time-distortion factor is very large, but finite. Farther and farther from the event horizon, the time-distortion factor gets progressively smaller. We do not experience anything like this distortion in the gravitational environs of any ordinary star or planet. Only collapsars, or neutron stars nearly inside their gravitational radii, can cause such a pronounced effect on time.

In theory, this property of the black hole makes it possible to travel into the future, just as the great speeds we might attain in a powerful space vessel may allow time travel according to the principles of special relativity. Dismissing for a moment the fact that getting arbitrarily close to a black-hole event horizon and then escaping would require incredible energy, let us imagine what such a time journey would be like. Imagine that we were to visit a planet in orbit around a binary star system, one member of which is a black hole. We might see in that planet a good candidate for the future development of earth-like life forms: There might be oceans, rainstorms, lightning and plentiful amino acids. We might want to travel to the distant future to see how the planet evolves.

The sober preparations for a no-return time voyage of a half billion years would involve careful checking of the size and location of the singularity, and precise calculations of engine thrust and the necessary depth of penetration into the gravity field. Then we would depart, descend to within a small distance of the event horizon, and return. At the closest approach to the singularity, the black horizon of the collapsar field would fold up ominously around us; only a tiny point of light would be visible from the outside universe. Then we would fire our escape engines, and our vessel would begin accelerating toward that tiny point of light. At that moment, we would doubtless be praying for the accuracy of our calculations, and our eyes would be fixed on the dot of light at the end of the time tunnel. Then the point would begin to grow wider, and individual stars would once again become visible; the stars would rapidly spread over the whole sky; the black hole would recede behind us, and then it would seem to be only a harmless little black dot in the heavens. What would await us down on the planet?

BLACK-HOLE UNIVERSES

The preceding examples of space and time travel involving black holes are a little far-fetched, at least by today's way of thinking. We have a lot of technological work to do before such escapades can even be seriously considered. Until then, about all we can do is speculate about the use of black holes in space and time journeys. Still, the time may someday come.

As far as venturing inside a black hole is concerned, the necessary technology is much simpler. All we would have to do is let ourselves be sucked inside the event horizon! But, such an excursion would be pointless for the purpose of gathering any information for the rest of mankind. No data could ever be sent back to the outside world; all radio information would be trapped inside the singularity with us.

A moderate-size black hole, harboring a neutron star, or a larger singularity, perhaps containing stellar matter of ordinary density, would be impenetrable for all practical purposes (if entering any black hole could be considered "practical"). The gravitational forces would crush us as we landed on the surface, supposing we would find a surface. But, a huge black hole might have a low enough density to allow us to fall freely within the event horizon without fear of death. What might we then find?

During our descent, the black rim of the sin-

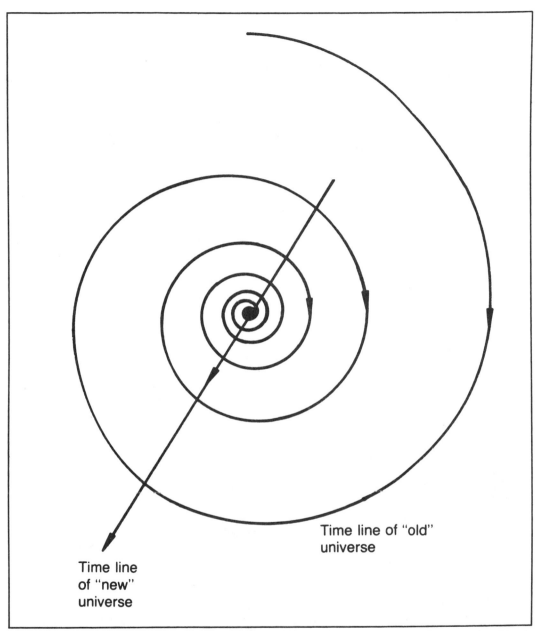

Time line of "old" universe

Time line of "new" universe

Fig. 8-10. Geometric representation of the time transfer in a black-hole singularity. The spiral shows the progression of time in the "old" universe (prior to entering the black hole). Arrows denote the direction of the future. The spiral gets half as large with each revolution, thus continuing infinitely toward the center but never reaching it. Jumping into the event horizon is represented by a direct approach to the center point. The time line in the "new" universe goes off in an entirely different direction (straight line).

gularity would seem to grow larger and larger, ultimately enveloping us. The outside world would retreat to a tiny point of light behind us, and then disappear altogether. At this moment, although we would be unable to observe the details, we would be haunted by the knowledge that all eternity had passed in the universe we knew. Whatever the final fate of our world, it would have been met in that brief instant. There would be no return. An event horizon is thus a one-way membrane. We could look only forward; our world would be behind us forever. Observers from the outside watching our disappearance into the eternal, would see our ship frozen to the rim of the event horizon as if it had come to a complete halt; they would imagine, perhaps, that we had gone into permanent suspended animation.

Continuing on into the singularity, we do not know exactly what we might find. We don't have much of an idea of the fate of a collapsar after it has shrunk within its gravitational radius. It may simply disappear, being crushed utterly out of existence. If so, then we would be doomed along with it. Such total annihilation seems intuitively unappealing; our journey into any black hole would be sheer suicide under such conditions. It seems that the matter must go somewhere—but where?

Perhaps the matter inside a black hole passes through the zero-volume point and reappears in some other form, beyond eternity. Or perhaps there is some outward force that balances the inward pull of gravitation at some time before the object can crush itself out of existence. This latter possibility makes our speculation most interesting, since it would imply that our bodies need not be squeezed into, or through, a geometric point of infinitely small dimensions. Let us imagine, then, our journey in this case.

Looking ahead, we might see a new point of light come into view. It would seem to grow as we fell into the new universe. We might find a small universe similar to our own: finite but unbounded, having three spatial dimensions. Individual stars might well survive a collapse into a huge singularity of low density. Planets orbiting such stars might be ripped away from their suns because of gravitational interaction, but some planetary orbits might

remain essentially intact; some planets might fall into orbit around new suns. We would have to begin a search for a new home planet without delay, since returning to earth would be out of the question.

Behind us would be a singularity, very similar to that we entered in leaving our old universe—the event horizon through which we had just come. We might see a new star emerge from this black hole now and then. Having ventured too close to the event horizon in the old universe, it, like ourselves, would have been irrevocably committed to the new.

Our knowledge of black holes should not tempt us to dive back into the singularity with any expectation of returning to the old world. We would have to accept that that universe was gone forever, existing as it were on a totally different time line from our new world. The old universe would have lived and died over the infinitude of ages. We might imagine this as shown in Fig. 8-10. The time-line of our old universe might be represented as a spiral which becomes half as large with each rotation, the future being inward toward the center. An infinite number of rotations would be possible in such a spiral, representing progression of the years infinitely into the hereafter. Entering a space-time event horizon would be like jumping all the way to the center of the spiral and then continuing outward along the perpendicular line. This straight line represents time in the new universe.

Failing to find a suitable habitat in the new universe, we could take the chance of diving once again into the singularity, but we could at best hope to emerge into yet a third universe, infinitely removed in time from the second! We would run the risk, too, of falling into a hostile world; each descent into a black hole would be a gamble, perhaps for the better, perhaps for the worse, and perhaps for oblivion. Even if such journeys ever become possible, men might never dare attempt them.

IS OUR UNIVERSE A BLACK HOLE?

If black holes indeed lead into other universes, then it is possible that our own universe is a black hole, itself attached to some external world by a space-time singularity. An external universe, with respect to which our world is a black hole, would exist

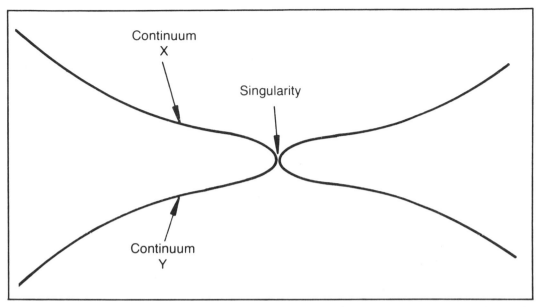

Fig. 8-11. Singularity between two universes. Matter might be sucked out of continuum X into continuum Y through the singularity; then it appears as a black hole in X and a white hole in Y. It would be impossible to turn back around, go through the singularity, and re-enter the other universe again, since an infinite period of time would have passed on the first journey through the singularity.

in the infinite past by our reckoning. The singularity connecting our universe with another would appear as a "white hole," spewing out material captured by the black hole from the other universe (Fig. 8-11).

One theory of the structure of our universe postulates that our entire cosmos is one great black hole. Observations of the density of matter, the known extent of our universe, and the known total mass therein, have led to the suggestion that our entire space is at or within its event horizon! Opponents of such a theory would object by pointing out that if this were true, then we should expect that the universe is undergoing a gravitational collapse, and is contracting. Observed evidence contradicts this: Our universe appears to be getting bigger, not smaller, as time passes.

We may now leave the question of black-hole "mini-worlds" and entertain the riddle of the origin, structure, and future of our own universe. Many theories about the nature of the cosmos have been put forth, beginning with the earliest men and progressing from philosophical dreaming through theological dictum to scientific inquiry. Modern astronomers pretty much agree on a cosmology that has come to be known as the "big-bang" theory. It got this strange name from the evidence that all the material in space was once concentrated in a single place, and then rapidly expanded in all directions. This outward motion is evidently still going on. Other theories to explain the expansion of the cosmos, or which deny that it is actually expanding at all, have gained less support. Let us now look at the science of cosmology, as it has evolved over the ages.

Chapter 9

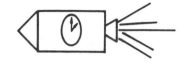

The Structure and Evolution of the Universe

FFROM THE TIME OF THE EARLIEST MEN, THE-
ories have existed concerning the nature of the
heavens. The word "universe" means "all that is."
Surely man has always been tantalized by the riddle: What is the structure of the physical universe?
Where did it come from? What is going to happen to
it? The physicist and the theologian share no more
common ground than the sciences of cosmology,
which is the study of the nature of the universe, and
cosmogony, which involves its origin and evolution.
This interrelation between the Church and the scientist has not always been very cordial.

What thoughts went through the minds of
primitive men as they gazed at the nighttime sky?
You can get some idea of their feelings by making a
trip to the country yourself, well away from man-
made light sources, and looking skyward on a
cloudless, moonless night. Under such ideal conditions, many thousands of individual stars can be
seen; the hazy band of the Milky Way has millions
upon millions more visible (although not individu-
ally discernible) stars. You get a feeling, by ob-

serving stars of various brilliances, of great depth.
Perhaps some primitives realized that not all the
stars are equally far away from the earth. But evidently most ancient philosophers and scientists
thought of the stars as affixed to one great solid
sphere around the central planet earth.

Even the earliest societies saw that the sun
traversed the heavens with great regularity. Sundials were the earliest clocks. The stars seem to
revolve around the earth at about the same rate as
the sun; but ancient astronomers noticed a change
in star patterns from season to season, and deduced
that the sidereal (star) day is just a little shorter
than the synodic (sun) day. Obviously, however,
both the sun and the stars could be plainly seen to
revolve around the earth. Obviously, since there
was no sensation of motion on the earth, the planet
was stationary, and thus it must be the sun and stars
that moved. Naturally enough, the earliest cosmology put the earth at the center of a ceaselessly
revolving universe. (From the general theory of
relativity, we can actually construct this cosmology

using the appropriate non-Euclidean coordinate system.)

In the most primitive models of the cosmos, the stars were attached to a distant solid sphere which completed one rotation each 23 hours and 56 minutes. The sun and moon were attached to spheres nearer the earth. The sun sphere had a period of 24 hours and the moon sphere had a period of about 24 hours and 50 minutes. But, there were still other objects in the sky besides the moon, the

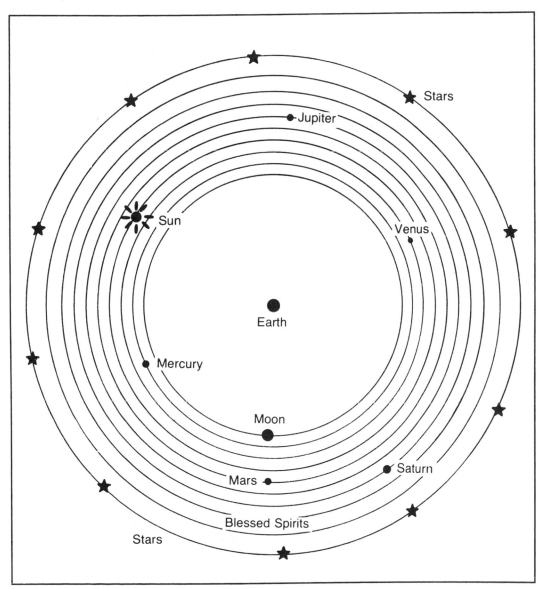

Fig. 9-1. The first model of the universe. The earth was at the center, and all the stars, planets, and other celestial objects revolved around it. In the far reaches of space, the blessed spirits supposedly reigned—the controllers of the great machine.

sun and the stars. These objects appeared at first to be stars; they were points of light just like the stars. Careful observation over a period of days or weeks, however, revealed that these strange objects changed position. They were called "wandering stars" or "planets." To the unaided eye, five such objects are visible. They were named after pagan gods of the time: Mercury, the messenger; Venus, the goddess of love; Mars, the god of war; Jupiter, the greatest god; and Saturn, the father of Jupiter. The planets were assigned geocentric spheres among the sun and stars (Fig. 9-1).

As all these spheres turned at their independent rates in the heavens, some ancient philosophers theorized that they must rub together and create noise. This noise would of course be extremely faint, and only the most favored ears would be able to hear it. A few claimed to hear the celestial spheres as they turned, and they said it sounded like incredibly beautiful music. This could only be the sound of the angels and other heavenly host, orchestrating the harmonious operation of the universe!

THE UNIVERSE OF PTOLEMY

The model of Fig. 9-1 was soon seen to be too simple to accurately explain the motions of the planets through space. According to the cosmology shown in Fig. 9-1, the planets would have a constant and uniform motion with respect to the background of distant stars. Observations, however, revealed that the planets did not travel through the heavens at a constant pace. Instead, they were seen to occasionally slow down, stop, and then temporarily reverse their direction. Plotting the position of a planet over a period of several weeks with respect to the distant stars, we do not obtain a simple line, but instead a line with a loop in its path (Fig. 9-2). The second-century philosopher and astronomer Ptolemy was the first to provide an explanation for this periodic retrograde motion of the planets. The resulting theory is known as the Ptolemaic system, in recognition of him.

Rather than following a simple, circular orbit around the earth, the planets were instead assigned orbits consisting of two distinct and independent motions, called the "deferent" and the "epicycle" (Fig. 9-3). The planet itself was supposed to orbit about a point in space which in turn followed a constant, circular orbit around the earth. By choosing the proper ratio of epicycle to deferent size, the motion of any planet could be very closely approximated. However, small errors were found even with this sophisticated system. Additional epicycles were added to provide theoretical consistency with observed fact. The network of epicycles and sub-epicycles eventually became so complicated that skepticism, or at least cynicism, arose. One ancient king said that he might have given some advice at the time of creation if he had had the opportunity!

The ancient geocentric theory of the universe was tailored to fit the observed facts, without acknowledging that perhaps the entire basis for cosmology ought to be rehashed. It was just too much for man to swallow; how could he, with all his magnificent edifices and almighty gods, be given any other place than the center of the universe? If not earth, then what could occupy such a cherished position? Even today, we sometimes see cases in science where emotion overrides rationality. The ancients preferred to complicate matters without limit, rather than admit a fundamental error.

COPERNICUS, TYCHO, AND KEPLER

In 1543, Nicholas Copernicus published his cosmology, in which the earth was described as one of seven planets revolving about the sun. Other astronomers had suspected that this might be the case, but none had dared stand up to popular consensus and proclaim his theory loudly. There was too much fear of reprisals from the Establishment, which was firmly convinced of the truth of the Ptolemaic system. Copernicus was not able to conclusively prove his theory. How could the earth be in motion, and continue the same motion without end, unless it were being pushed around and around the sun? Would the earth not come to a stop without continual pushing? Since no great cosmic god could be seen manipulating the earth in a circle about the sun, the earth was assumed to be stationary. Copernicus, however, applied this same reasoning

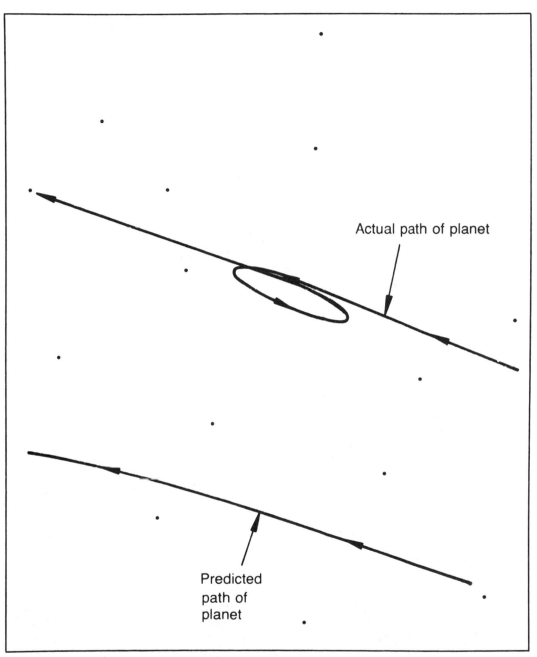

Actual path of planet

Predicted
path of
planet

Fig. 9-2. Observation shows that the model of Fig. 9-1 must be modified slightly. The actual path of a planet with respect to the background stars contains an occasional loop. The predicted path, however, should not contain such loops if Fig. 9-1 were indeed the correct representation of the universe.

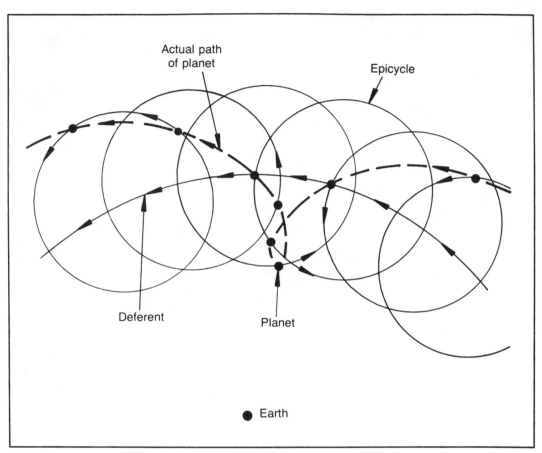

Fig. 9-3. The explanation used by Ptolemy to explain loops in planetary motion. The "deferent," or main orbit, of a planet was imagined to be a circle centered on the earth. The planet, however, did not exactly follow this path, but instead orbited a point on this track. The smaller orbit was called an "epicycle." The actual path of the planet, then, would sometimes have a loop, as shown by the broken line. This explanation satisfied most scientists for a thousand years.

to the other planets: Why should they move in circles around the earth, the sun, or any other object, unless invisible forces were involved? Such forces, necessary to account for the motion of other objects in the universe, could also act on the earth, Copernicus reasoned.

The motions of the planets through space were recorded with great accuracy by Tycho Brahe during the 16th century. Tycho's assistant in his later years was a man named Johannes Kepler. Kepler formulated a theory based on the postulates of Copernicus and the observations of Tycho; the re-

sult became popularly known in the form of Kepler's Laws, published early in the 17th Century. Kepler deduced that the planets do in fact orbit around the sun, but their orbits are not perfect circles. Instead, the orbits of the planets are elliptical, having varying degrees of eccentricity, with the sun at one focus. Kepler showed that the planets move faster when they are near perihelion (the closest approach to the sun) as compared with the orbital speed at aphelion (the most distant point from the sun). Finally, Kepler found that the period of revolution of a planet depends on its distance from the sun. Letting the period be T and the distance be d, then:

$$T = k \sqrt{d^3}$$

This is where k is a constant that depends on the units of time and distance. These laws of Kepler are illustrated in Fig. 9-4.

Many scientists dismissed the propositions of Copernicus and Kepler, along with the theologians, as heresy. The Establishment went on believing the cosmology of Ptolemy, which had endured for over a thousand years. The upstarts were not to be taken seriously. More convincing evidence was needed to verify the truth of the heliocentric (sun-centered) theory of the universe, even though this new theory was much less complicated than the old one, and explained the retrograde loops in the ce-

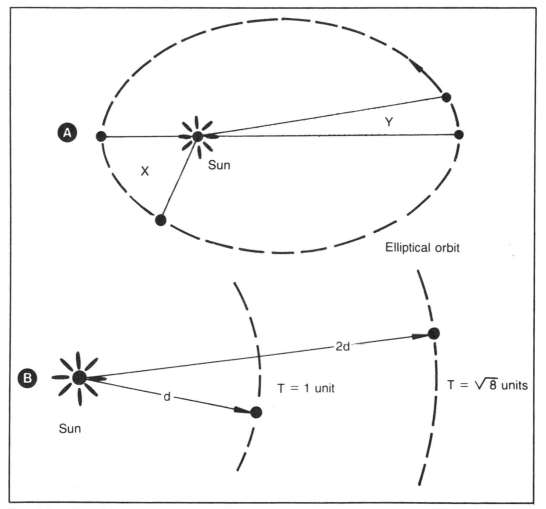

Fig. 9-4. According to Johannes Kepler, the orbits of the planets were ellipses with the sun located at one focus (A). The speed of each planet around the sun was such that equal areas such as X and Y would be traversed in equal lengths of time by a line connecting the planet with the sun. The revolution period of a planet was shown to be dependent on its distance from the sun. Doubling the distance would increase the period by a factor or $\sqrt{2^3}$, or $\sqrt{8}$, as shown at B.

lestial paths of the planets with much greater accuracy.

GALILEO AND NEWTON

An Italian astronomer named Galileo Galilei, a contemporary of Johannes Kepler, also believed that the earth revolved around the sun. Galileo used the first astronomical telescopes, and discovered mountains on the moon and satellites revolving around the planet Jupiter. He also was responsible for the discovery that objects of different mass fall with equal speed in the earth's gravitational field. The Establishment took Galileo seriously: He represented a grave threat to their theories. The man invited skeptics to look through his telescopes for themselves; he showed people how the speed of free fall is independent of weight. Galileo was confined to his home for the last few years of his life.

The heliocentric theory gradually gained universal acceptance after the work of Sir Isaac Newton of England. Newton provided the first logical explanation for the invisible forces that keep the planets in constant motion around the sun, and keep moons in orbit around their planets. It is sometimes said that Newton gained inspiration by watching an apple fall from a tree in his garden. (Some stories even say that the apple hit Newton on the head!) Whatever the actual impetus, Newton noted that the earth exerts a constant attractive force on all things in its vicinity. All objects also exert an attractive force on the earth, and on one another. Newton condensed the work of his predecessors, as well as the results of his own research, into his Laws of Motion, published in *Principia* in 1687. This threefold theory maintains that (1) Any object at rest will stay at rest unless some force acts on it, and any object having uniform motion will continue that motion unless an outside force intervenes; (2) A force on any object causes an acceleration in the exact direction of the force, in direct proportion to the force, and in inverse proportion to the mass of the object; and (3) Every action results in an equal and opposite reaction. These laws remain the basis for classical mechanics to this day. Except when extreme forces, masses or velocities are involved, Newtonian mechanics explains observed phenomena with almost perfect accuracy.

In order to explain the acceleration of falling objects, and the orbits of the moon and planets, Newton proposed his principle of gravitation: Every object attracts every other object in the universe. For two objects X and Y, the force of attraction is proportional to the product of their masses $m_X m_Y$, and also to the inverse square of the distance $d = XY$. This is shown in Fig. 9-5 for two specialized cases.

At last it became evident how the planets could orbit the sun. They are all actually accelerating toward the sun because of gravitation, but this inward force is exactly balanced by the inertial resistance of each planet to changes in direction. Every planet is in a state of free fall about the sun. The moon is falling around the earth; Jupiter's moons are falling around that planet. The origin of the mysterious force called gravity was not clear, but its existence was.

ORBITAL GEOMETRY ACCORDING TO NEWTON

Johannes Kepler showed that the orbits of the planets are not perfect circles. In fact, it is a rare coincidence for the orbit of any object to be circular; there is almost always some deviation from that state of perfection. The moon changes in distance from the earth on its monthly circumnavigation of our planet. The earth is about two million miles nearer the sun in January than it is in July. Some asteroids have highly eccentric orbits, elongated so that they do not nearly resemble circles. Some celestial objects are pulled into the solar system by the gravitation of the sun, make one near pass, and then depart never to return.

The path of an object from outside the solar system is, of course, not an ellipse, but an open path. Newton showed, though, that all orbital paths, whether closed or not, have the geometry of a *conic section*. A conic section may be a circle, ellipse, parabola or hyperbola. Figure 9-6 illustrates examples of these four possible orbital configurations.

You can easily generate these conic sections using a bright flashlight on a large, smooth surface in a dark room or on a dark night. Standing on the

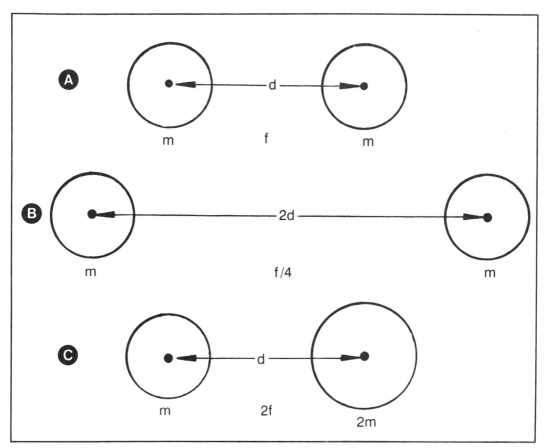

Fig. 9-5. Newton's principle of gravitation. Two objects of mass m, separated by distance d, might exert a force f of mutual attraction (A). Doubling the distance (B) cuts the attraction to one-quarter its previous value; doubling one mass (C) while leaving the distance constant will result in a doubling of the mutual attractive force.

flat plane of, say, a tennis court or parking lot, shine the flashlight straight down. Note the perimeter of the dim, wide-angled portion of the flashlight beam (not the bright central part); it forms a circle. By tilting the flashlight, you can make the circle become elongated; this is an ellipse. The more you tilt the flashlight, the greater the eccentricity of the ellipse. At a certain tilt, the ellipse becomes so elongated that the far end disappears from the plane surface; the edge of the light then is a parabola. If you tilt the flashlight until it is shining horizontally, the edge of its wide light cone lands on the ground in the form of a hyperbola.

Given an object in perfect circular orbit around the sun, Newton realized that, if it were given a slight nudge, its orbit would become distorted into an ellipse; a larger nudge and the orbit would take the object away from the sun forever, following the curve of a parabola. A very forceful push, much more than the minimum needed to throw the object permanently out of the sun's gravitational proximity, would cause the object to follow the curve of a hyperbola. All these kinds of orbital situations occur in space. (The parabola and hyperbola are probably the most common paths for space objects, since there are a lot of "free floaters" out there!)

The mutual attraction of all objects in space results in some displacement of the earth as the moon revolves, and some displacement of the sun as the earth revolves. Two equally massive objects would, if placed in proximity, orbit each other around a common center (Fig. 9-7A). In the case of two masses of unequal size, the common center of their mutual orbit is closer to the larger mass (B); but never does one mass stay totally motionless as the other orbits it. Thus, according to the New-

tonian theory, even the tiniest manmade satellite causes displacement of the earth as it revolves. Although Sputnik or Spacelab did not create significant displacement of the earth, the moon certainly does. The common center of the earth-moon mutual orbit lies 2,900 miles from the earth's center.

NEWTON'S ABSOLUTE SPACE AND TIME

The theory of gravitation, developed by Newton to

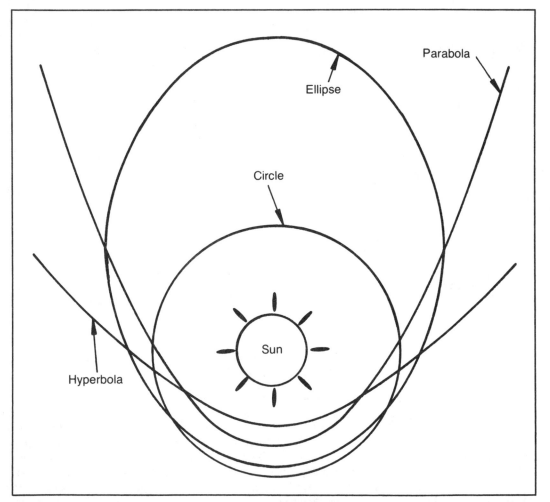

Fig. 9-6. Various possible orbits around the sun. All are conic sections. Objects that stay in permanent orbits follow either a circular or elliptical path. Temporary visitors to the solar system follow parabolic or hyperbolic orbits, and never return.

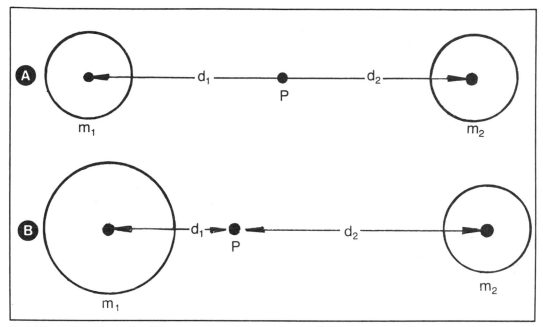

Fig. 9-7. Mutually orbiting bodies. If the masses m_1 and m_2 are the same (A), then the center P of the orbit lies midway between the centers of the objects, so that $d_1 = d_2$. But, if $m_1 > m_2$, then $d_2 > d_1$ (B). Whatever the ratio of m_1 to m_2, however, the center point P is always at least a little bit displaced from the center of the heavier object.

explain the motions of the planets and their satellites, at last solved the mystery of the circular and elliptical orbits, and the reason for their ceaselessness. But, what implications about the greater cosmos might be derived from this powerful theory? If every object attracts every other, even at extreme distances, the structure of the whole universe might be significantly affected.

The first attempts to measure the distances to the stars were carried out by the parallax method, with two observations made at widely separated points. The orbit of the earth has a diameter of some 186 million miles, and this distance was used as a base line for triangulation. When observed in, say, April, some stars appear slightly displaced relative to the background of much fainter stars, as compared with their positions six months later in October. It was immediately apparent that the stars were all very distant! But, it would seem that the combined gravitational effects of the billions upon billions—and possible infinite quantity—of stars

would be very large, even at great cosmic distances.

It appears that Newton failed to see the link between the omnipresent gravitational field, produced by all the matter in the universe, and the laws of inertia. Instead, Newton ascribed the existence of intertia to "absolute space." This is where Newton departed the empirical world for the purely theoretical, and his conclusions thus became speculative. Why does the moon not fall into the earth, or the earth into the sun? Evidently, the moon was revolving with respect to absolute space, and so was the earth. This notion of space as an entity possessing absolute properties was lent support by the Foucault pendulum experiment (Fig. 9-8).

In this experiment, constructed to prove that the earth rotates, a heavy weight was suspended by a fine wire over 200 feet long, and the weight was set swinging. Care was exercised to ensure that air currents and other disturbances did not interfere with the precise plane of the pendulum's movement. To be certain that the pendulum was not disturbed at the outset, it was started in motion by burning a cord that held it perfectly still and to one side. As the hours passed, the plane of the pendulum was seen to rotate. This appeared to be quite satisfactory proof that the earth rotates with respect to absolute space.

A Foucault pendulum at the equator will not rotate at all, since the plane of its motion always contains the celestial poles (Fig. 9-9A). At either of the poles, the pendulum's plane of motion will rotate once every 23 hours and 56 minutes, keeping its original orientation with respect to the background stars (B). At intermediate latitudes, the plane of motion rotates more slowly than at the poles, depending on the distance from the equator (C). To the classical physicist of Newton's time, and up till Einstein's general theory of relativity became widely known, this effect could be ascribed to only one thing: The plane of the Foucault pendulum rotates because of the effects of absolute space.

From this concept, Newton germinated the idea of absolute time as well. According to Newton, "true" time flowed smoothly, unaffected by anything external. This was indeed a bold assumption. While it is intuitively pleasing, it is based on essentially no observed facts. Relativistic physics proves that absolute time, as well as absolute space, are illusory.

In the generations following the acceptance of

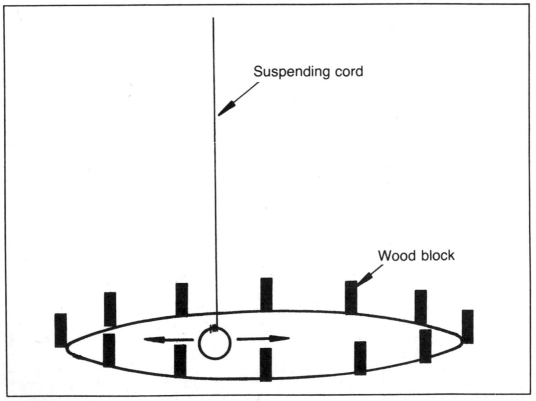

Fig. 9-8. The Foucault pendulum experiment. The cord must be very long, and there must be no air currents to disturb the swing of the weight. As the plane of the motion rotates, the weight will knock over all of the wooden blocks after a length of time, except at the equator.

184

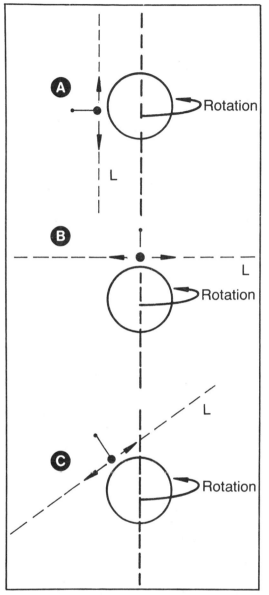

Fig. 9-9. At A, a Foucault pendulum at the equator does not rotate because its plane of motion lies in line with the celestial poles. At B, a Foucault pendulum at the pole turns "under" the earth; at C, a Foucault pendulum at intermediate latitudes rotates more slowly than at the pole. The celestial axis of the earth is shown by the heavy broken line. This experiment was believed to prove that absolute space existed, until Einstein showed that such a notion was false.

Newton's laws of gravitation and motion, many observations were made by astronomers as methods of measurement became more and more refined. Would Newton's theories accurately predict the celestial motions of all objects? Or would some exception be found? By that time, inquisitiveness was no longer the mark of the heretic; science became more research-oriented as man strove in greater unison to find the truths of the universe. Newton's laws seemed to accurately predict observed phenomena within the limits of measurement accuracy, with one nagging exception. But, one contradiction is as good as a thousand in the realm of logic.

The perihelion of Mercury changes position by 43 seconds of arc per century. Newton's laws could not explain this. The astronomer Leverrier, in 1845, carefully observed the precession of the orbit of Mercury, and it was definitely real. There it was: the one indisputable flaw in an otherwise apparently perfect model of the universe. Attempts were made to explain this strange observation, but a satisfactorily model did not come into being until Einstein's time.

THE ADVENT OF MODERN COSMOLOGY

The beginnings of relativity theory, which set the minds of astronomers in a new direction, were conceived when scientists began to seriously investigate the nature of radiant energy. The speed of light was discovered to be finite. Newton has proposed that light was like a barrage of particles. Particle-like properties of light were indeed found, but some experimenters noticed that it behaved like an oscillatory, or wave-like, effect. It was natural to ask why light traveled through interstellar space; evidently there was no air in those far reaches, since orbiting bodies do not slow down from friction. How could any effect be propagated through a perfect vacuum? Newton's particle theory explained the problem tolerably well, but observed diffration effects indicated that light had oscillatory properties. What was doing the oscillating?

This question led to the postulate that some sort of propagation medium, called the "lumeniferous ether," existed for electromagnetic waves, just as air and water are conductors of sound, and wires carry electricity. Curiously, though, this ether did not seem to affect the motions of heavenly bodies, and apparently had no mass. Great efforts were made to find the ether and demonstrate its properties more rigorously. Here is a perfect example of a fabrication for the sole purpose of making things intuitively agreeable; it is a dangerous practice in any logical pursuit of knowledge. We have seen (Chapter 1) that the ether theory for light propagation was ultimately discarded. This brings us to the time of Einstein, who postulated that the speed of light is simply a manifestation of space and time, independent of the viewpoint of an observer. This got rid of the need for the ether theory.

Early in his life, Einstein is said to have asked himself what one ought to see if one could travel alongside a beam of light, keeping pace with it. He believed that if he could do such a thing, he would see a spatially oscillating electromagnetic field at rest. It was this answer that eventually led Einstein to formulate the special theory of relativity, disproving the Newtonian idea of absolute space and time. Later, Einstein generalized this theory to show that all reference frames, even accelerating ones, are equivalent to all other points of view if we use suitable transformations. The general theory of relativity explained the precession in the orbit of the planet Mercury. It accurately predicted the bending of light by the gravitational field of the sun. It accurately predicted blue and red shifts of radiant energy in the gravitational influence of the earth. And, the theory aroused curiosity about the structure of the whole universe. For the first time, meaningful results were produced concerning this strange and great question.

The distortion of space in the vicinity of a source of gravitation (and that means anything that has mass) extends in all directions without limit. Even at a distance of millions of miles, the gravitational influence of a single atomic nucleus contributes to the curvature of space. There are trillions upon trillions of atomic nuclei, of course, scattered all through space. While they tend to be concentrated here and there in the form of stars, planets, planetary satellites, asteroids, comets and other solid objects, there are atoms even in the remotest parts of space. Einstein realized that each and every one of these atoms would have an effect on the geometry of the entire cosmos. His equations led him to the conclusion that space must be finite yet unbounded: The universe is literally curved all the way around on itself in all directions! Assuming that the density of matter, averaged on a large scale, is about the same everywhere in the universe, Einstein concluded that the universe must actually be the three-dimensional surface of a four-dimensional sphere. Thus, there should be a finite number of atoms in the cosmos (although it is surely a very large number), but no boundary. The layman initially reacts to this proposition with disbelief—the same disbelief that was at first directed toward those who dared to say the earth was not flat, but round.

The single apparent flaw in this theory seemed to be that, if the amount of matter in the universe is not infinite, then its overall gravitational attraction ought to cause it to collapse inward. Figure 9-10 illustrates this. Reducing the universe by one geometric dimension for ease of visualization, we can imagine that an attractive force among all the atoms in space should cause contraction of the cosmos. This was received as an objectionable idea; where would the matter then have originated? Einstein countered by postulating that a repulsive force, acting inside the four-sphere, kept the collapse from happening. The gravitational attraction should then exactly balance the repulsive force at a certain radius. If the universe were smaller than this size, it would expand; if larger, it would contract to this equilibrium point.

But, then again, perhaps the universe really could be getting smaller. How would we be able to tell? The answer lay in the spectra of distant objects. In a static universe—one that has no change in size with the passage of time—no spectral shift would be seen, except that caused by random motions. In a contracting cosmos, however, distant objects would display a spectral shift toward the blue, and this shift would appear greater with increasing remoteness. And there was a third possibility, too: A spectral red shift might be observed,

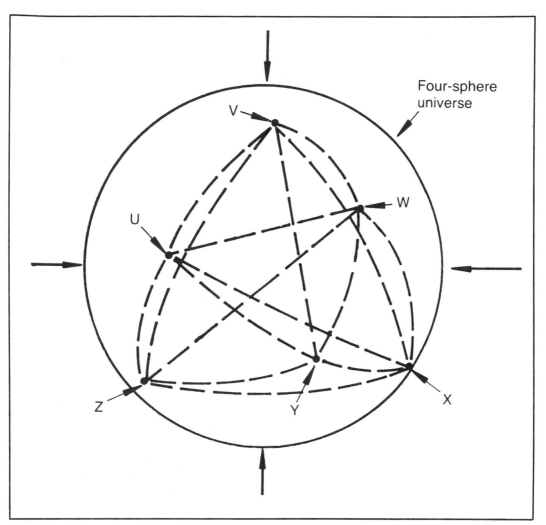

Fig. 9-10. Mutual attraction of six objects on the surface of the four-sphere universe (here reduced by one dimension for visualization). The broken lines show the lines along which the gravitational attraction acts among the objects U, V, W, X, Y and Z. This effect, among all atoms in the cosmos, was at first thought to mean that the universe must be contracting (heavy arrows).

in greater magnitude with increasing distance. This would mean that the size of the universe was becoming greater with time. Before such observations could be made with any authority, accurate methods had to be developed for determining the distances to celestial objects too far away to be triangulated.

DISTANCE MEASUREMENTS IN THE COSMOS

Einstein's first estimate of the diameter of the four-sphere universe was approximately 100 million light years (about a thousand times the diameter of our galaxy, although the size of the Milky Way was not yet known). The four-sphere model of the universe has gained more and more acceptance since Einstein first put the idea forth in the early part of this century; however, estimates of its overall size have been revised upwards. Not long after Einstein first proposed that space is finite but unbounded, it was discovered that certain nebulous patches, which had been thought relatively nearby, were really very far away.

One of the most useful tools for measuring the distance to remote objects makes use of a peculiar kind of star called a Cepheid variable. Named after the constellation Cepheus, where the first such star was found, such objects display a regular periodic change in their brightness, ranging from a day or two to several weeks in duration. The fluctuations are significant enough to allow the period to be very closely measured. Cepheid variables are bright stars, and they can be seen from extreme distances. Harlow Shapley noticed in 1917 that the period of the luminosity change seemed to depend on the brightness. If this were true, then it would be possible to determine the distances to remote star clusters containing at least one Cepheid variable. Shapley developed a relation for the absolute brightness as a function of the period. This function has been revised several times since, but it allows fairly accurate determination of the distance.

There is some margin for error in measuring cosmic distance using Cepheid variable stars. First, it is not always easy to know how much interstellar dust lies between us and a particular star; this dust might have a large effect on the observed brilliance. Second, the absolute magnitude of a Cepheid variable depends on other factors besides its period. The temperature of the star's surface, for example, has some effect, independent of the absolute brilliance. But, when some nebulous patches containing Cepheids were carefully observed, the resulting estimates of their distances were astounding. These nebulae were evidently ouside our own Milky Way, and not just a little outside, but very far away. All the stars in our galaxy appeared much nearer. Careful photographing of these strange nebulae revealed fantastic geometric configurations: spheres, ellipses, spirals, and irregular shapes. The existence of other galaxies had been discovered. The distribution of stars in space was found not to be uniform, as had been thought. Instead, the stars were in vast congregations of hundreds of billions apiece! The galaxies themselves were found to occur in clusters, ranging from a few to thousands.

The most distant visible galaxies appear, according to most recent estimates, to be billions of light years away. But, could it be that we are looking all the way around the four-sphere universe, perhaps several times (Fig. 9-11), and that the actual circumference of the universe is much less than those billions of light years? Could it be that one of those far-off spiral galaxies is our own Milky Way as it appeared 300 million or more years ago? Astronomers cannot positively answer this, but the modern consensus is that we have not yet seen all of the universe. Current belief holds that the diameter of the four-sphere cosmos is at least 10 billion, and possibly 100 billion or more, light years. In the future, more accurate observations and further data should give us a better idea of the full extent of the cosmos.

SLIPHER, DE SITTER, AND HUBBLE

The first astronomer to notice that some galaxies seemed to be moving away from the earth was Vesto Melvin Slipher. He found spectral red shifts in the light from several distant galaxies. These red shifts apparently indicated that the galaxies were retreating at a very high rate of speed. Actually, Slipher's discovery was an accident. He was not even looking for this result. But, he announced it at a meeting of astronomers in 1914. The discovery was received with unprecedented enthusiasm in the scientific community; cosmologists did not know exactly what it meant, but they had a feeling it was extremely significant. Their hunch has proved to be well founded.

At about the same time Slipher made his accidental finding, a Dutch astronomer named Willem deSitter found that, according to the equations of Einstein's general theory of relativity, the size of the universe ought to be constantly increasing. Another scientist, the Russian mathematician Alexander Friedmann, found a different solution to

Einstein's equations that also predicted an expanding universe. At first, Einstein did not take this seriously, perhaps because of his own emotional preference for a static cosmos. However, Einstein was finally convinced that he had made a mathematical error in his own calculations that led him to believe the universe was unchanging in size. The error was a simple matter of inadvertently dividing by zero. But the notion of an expanding universe repulsed Einstein. He, and many other scientists, objected to the theological implications of such a theory.

To accept the idea of an expanding universe seemed to mean that there must have been a "be-ginning of time" when all the material in the cosmos was concentrated in one place. This would imply that, prior to the instant the expansion began, we should never be able to tell what happened. How would we ever determine the events before the beginning of all things? While some cosmologists have been upset by this strange impasse, others are awed, and some even point to this finding as proof of the existence of a supreme being—the only possible reason for the original formation of matter!

After Slipher first noticed the spectral red shifts of the distant spiral nebulae, he decided that further investigation should be conducted. He clocked the apparent velocities of many more

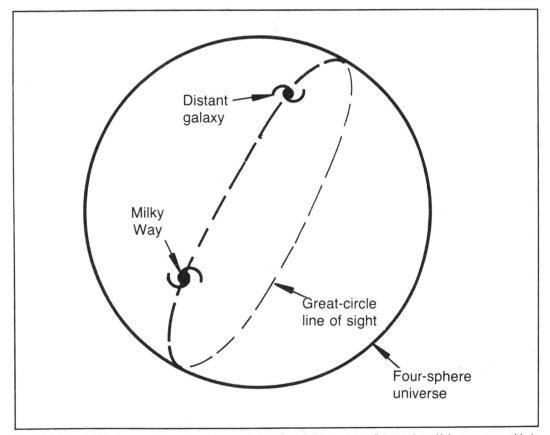

Fig. 9-11. We might look several times around a great-circle line of sight to see a distant galaxy. Unless we were able to conclusively show that the multiple objects were really the same galaxy, we might think the universe was much larger than it actually is.

galaxies, and found that nearly all of them were moving away from the Milky Way at enormous speeds. It was Edwin Hubble who compiled this data, along with the results of his own research, and came to one of the most famous conclusions that cosmology has ever known.

At first, no one seemed to notice the connection between Slipher's observations and the predictions of Willem de Sitter. Together with Milton Humason, Hubble began an extensive investigation of the distant spiral nebulae using the 100-inch telescope at Mount Wilson. Like Slipher, Hubble and Humason found red shifts in the spectra of these objects. All the very distant galaxies, they found, are moving away from the earth. Some are moving at greater speeds than others; some are receding with almost unbelievable rapidity. But, it was not immediately clear how distant the spiral nebulae in fact are. Some astronomers believed they were relatively small objects, within or near our own Milky Way system, and perhaps the red shifts occurred because the nebulae were being ejected from our own galaxy. Other scientists believed that the spiral nebulae were indeed "island universes," separated from our galaxy and from each other by enormous distances, and that our Milky Way was but one such system. Hubble found that all the spiral nebulae contained an enormous number of stars each; he measured the distances to some of them by means of Cepheid variable stars, as we have seen, and found that they were very far away. For galaxies too far away to be located using the Cepheids, Hubble estimated the distance by means of the combined brilliance of all the stars in the whole galaxy. Some of the galaxies were as far away as seven million light years! Since Hubble's time, astronomers have found objects that are billions of light years from us.

Hubble was the first to notice a strange correlation between the distances to the spiral nebulae, and their red shifts. It appeared to him that the farther away a galaxy was, the greater was its red shift, and thus the faster it must be moving. By making meticulous observations over a long period of time, he found the relation that we now call Hubble's Law of the expanding universe. The speed of a galaxy is a direct, linear function of its distance from us (Fig. 9-12).

Hubble's Law is actually a necessary mathematical result of a uniformly expanding universe. If the distances between all of the galaxies in the universe doubles in a certain period of time, then a galaxy at distance d should seem to be retreating at just half the speed of a galaxy at distance 2d, when the distances are measured at the same time (Fig. 9-13A).

You can observe Hubble's Law in operation by blowing up a spherical party balloon. Inflate the balloon somewhat, and have a friend make several dots on the surface of the balloon with a felt marker. Then, blow up the balloon to twice its former size. Every mark will then be twice as far from its neighbors as it was before, as measured over the surface of the balloon. This is illustrated by Fig. 9-13B. Two dots that were originally one inch apart will be two inches apart after the additional inflation; two dots that were two inches apart will become separated by four inches. As the balloon is blown up, the pair of dots originally an inch from each other will move apart at a certain speed v; the pair of dots originally two inches apart will move away from each other by 2v. This is the essence of Hubble's Law as it applies to galaxies in our universe. Apparently the four-sphere of the cosmos is expanding, just as the balloon expands when you inflate it. The galaxies are carried along in this expansion just as are the little dots on the balloon.

The slope of the line in the graph of Fig. 9-12 has been revised several times since the concept of the uniformly expanding universe was introduced by Hubble. The revisions are the result of improvements in the methods of estimating the distances to remote galaxies. Figure 9-12 shows the latest data.

Perhaps it is hasty to conclude that the red shifts in the spectra of distant galaxies are the result of motion. We have seen that a gravitational field can cause red shifts in the frequencies of radiant energy. Is it possible that the universe is not expanding, and that the red shifts we see are caused by something other than motion? Yes, this is possible.

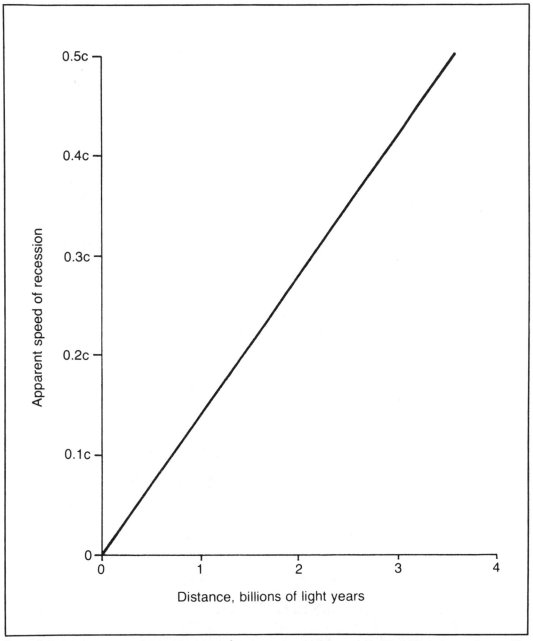

Fig. 9-12. The most recently updated data for the speed of recession of galaxies as a function of their distance from us. The slope of this line, approximately 0.12c per billion light years, is called the Hubble Constant, after the astronomer who first found the linear function.

Light beams traveling over the surface of a four-sphere universe are, in effect, orbiting a huge black hole comprised of all the matter in the cosmos. This light energy is constantly trying to escape the overall gravitation of the universe, but is being held in orbit by that gravitation. This might produce a spectral shift toward the red, and it would be expected to increase linearly with distance. Further research might provide better insight into this possibility. Red shifts may also be caused by other effects, such as a change in the rate of time progression as the universe ages. Light from the distant galaxies left them many millions of years ago; if the rate at which time "moves" is different now than it was then, we would see a shift in the spectra emitted by distant stars. Perhaps intergalactic matter is responsible for a loss in the energy of light propagated over great distances. This, too, could cause a red shift.

Arguments such as these have been proposed by astronomers who have difficulty accepting the idea of a "beginning of time." But, they are all quite tentative arguments. Other observations besides the red shifts in the spectra of distant galaxies lend support to the idea of an expanding universe. Before we examine the evidence that points to a cosmic explosion, or "big bang," billions of years ago, let us look at a theory that was invented for the purpose of avoiding a creation or beginning of time, but still explains how the universe is expanding.

THE STEADY-STATE THEORY

The density of matter in the cosmos appears, on the average, to be fairly constant, at least in the neighborhood of our galaxy. Determination of the density of the universe at great distances is complicated by the fact that it is hard to measure the separation of far-off objects with much precision. If the universe is expanding from a primordial ball of compact matter, however, we should notice a change in the density of the universe with increasing distance. If the cosmos was more tightly packed in the past than it is today, we should see greater and greater material density as we probe farther and farther into space. No conclusive data of this nature has been obtained, however, although there

is some evidence that the rate of expansion may have been greater in the past than presently.

This tiny thread of hope for those wishing to avoid a theory of massive creation has been used as an argument in support of an idea sometimes called the steady-state model of the universe.

A theory proposed by Hoyle, Bondi and Gold maintains that, while the universe is in fact getting bigger all the time, its density is held constant by the continuous generation of matter everywhere in space. This strange picture of the cosmos allows us to envision its expansion without changes in density, but its effectiveness in getting rid of the creation problem seems doubtful. Where would such newly formed matter come from? It supposedly forms, according to the proponents of the theory, at the incredibly slow rate of one atom per million cubic meters per century, and it forms, apparently, from nothing. The opponents of the creation concept are thus faced with exactly what they wish to avoid: Rather than one big event, there are billions upon billions of little ones. It seems absurd to assume that matter would form spontaneously at such a controlled rate; it would have to come from someplace, but we have no idea where. Furthermore, tracing backward in the steady-state model of the expanding universe leads to a time when there must have been nothing at all in the cosmos. Where did that first particle come from?

Variations of the steady-state theory have been proposed. Some astronomers have suggested that novae and supernovae explosions might be objects through which new matter is created. The same might be true of the quasars. Perhaps gravitational energy is converted, by some as-yet unknown cosmic process, into matter. Perhaps the entire universe is pulling matter in from another universe, as we discussed in Chapter 8. This last idea seems to be the best escape from the idea of creation; a huge collapsar might create a small universe within our cosmos, so why couldn't our universe be a great collapsing object in some other, external continuum?

The daydreaming can go on and on. Those who wish to believe in creation, in a beginning of all things, can find solace in the big-bang theory.

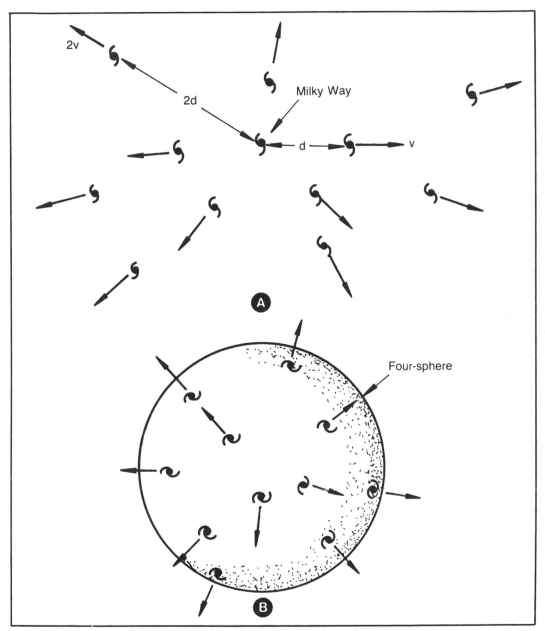

Fig. 9-13. Hubble's Law. At A, galaxies in the universe seem to retreat from us at a speed that depends on their distance. A galaxy at initial distance d may retreat at speed v; then a galaxy at initial distance 2d will retreat at a speed 2v. This would be the case as seen from any galaxy in the universe, not just our Milky Way. At B, the four-sphere universe expands, and carries the galaxies outward with it, creating the "spreading" effect we observe on the three-dimensional surface. It is very much like blowing up a spherical party balloon.

Those who dislike such an idea might prefer to believe that our universe is a huge black hole, expanding as it pulls matter in from somewhere else. More scientific investigation has been conducted, however, and the results seem to confirm the big-bang theory. Most modern cosmologists accept this model whether they "like it or not." Reason must, after all, override emotionalism in scientific inquiry.

THE BEGINNING OF TIME

The most powerful evidence for the big-bang theory was, like so many other things, discovered inadvertently. In 1965, Arno Penzias and Robert Wilson of the Bell Laboratories were testing a highly sensitive radio receiver, when they noticed strange background noise. The noise seemed to have no definite direction; it was coming from every region of the sky. The receiver was found to be in perfect working order; there were no faulty components that were causing the noise. It was coming from space!

This radiation had been predicted by other scientists as the cosmic birthmark of the big bang, but their results were ignored until the strange radiation was discovered by accident. No explanation has been found satisfactory to fit this phenomenon, except a primordial explosion billions of years ago.

According to the big-bang theory, all the matter in our universe was originally concentrated into a space smaller than a single atomic particle. Imagine the density of such an object; imagine how bright it must have been! Penzias and Wilson were evidently listening to the greatly red-shifted radiant energy that comes from that original fireball. It appears to arrive from all directions, because it is forever traveling around and around the three-dimensional surface of a four-sphere universe (Fig. 9-14). It will never totally disappear. It was present when stars first began to form from the whirlpools of gas in the galaxies; it was there when the earth was born; it was there when the dinosaurs perished. It is here at this moment, and it will be coursing through space when the sun burns out, and after every other star in the universe has gone cold.

Perhaps then, ironically, the original radiation from the birth of the cosmos will be the only evidence that anything ever took place there.

THE END OF TIME

If our universe began as a violent fireball, how will it end? Will the expansion carry the matter outward indefinitely, as in the gloomy scenario just described? Or is it possible that gravitation might pull it back together again?

This depends on how powerful the overall gravitation of the universe is, as compared to the speed with which the galaxies are separating. The total amount of mass, as estimated from the visible sources in the observable universe, seems inadequate to pull the universe back together again. But, as we have seen from the last chapter, much of the matter in space is probably not visible, occurring as intergalactic dust, gas, and dead stars and black holes. We do not know precisely how much of the matter in the cosmos is invisible, and this makes it hard to estimate the overall density of matter. Astronomers are constantly working on this problem, because it affects the outcome of cosmic events so profoundly. At present, it appears that there is not a great enough concentration of matter in the cosmos to stop the expansion. This would indicate a gradual end, rather than a violent collapse. It means that, apparently, the end will be cold and dark. In a few tens of billions of years, our universe will evidently have only lifeless chunks of matter: dust, gas, dead stars and collapsars. The primordial radiation will still be there, its energy being red-shifted more and more as eternity goes on.

If new discoveries are made that provide evidence for more matter in the universe, we may be able to offer a more optimistic prediction of the fate of our cosmos. It is possible that the galaxies might slow their rate of expansion to a halt, and then begin to fall back inward. This contraction would begin gradually, becoming more and more rapid over a period of billions of years. The stars and galaxies would have gone dark long since, but they would ultimately have a fiery end: The density of the universe would increase, and finally the matter

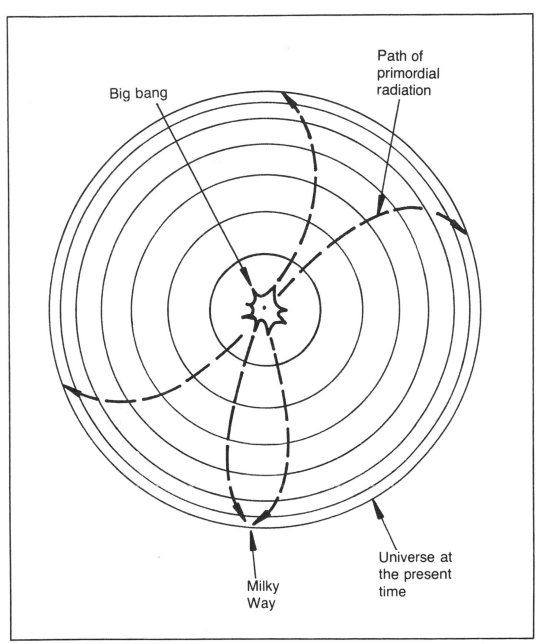

Big bang

Path of
primordial
radiation

Universe at
the present
time

Milky
Way

Fig. 9-14. We see the radiation from the primordial fireball coming from all directions. In this dimensionally reduced illustration, the paths of photons from that explosion follow curved paths through four-space as the universe expands. The rate of expansion is slowing, as indicated by the concentric circles which represent the universe at intervals of about two billion years. The most recent estimates of the age of the cosmos range from 10 to 20 billion years.

would be recompressed. Each atom would lose its identity. With all the force of the original explosion, the universe would be crushed down to an immeasurably tiny volume. It would appear as a reverse-motion picture of the primordial explosion.

It is sheer speculation to predict what might then happen. Time and space, reduced to a singularity as in the beginning, erases all clues as to what might take place before the birth, or after the death, of the cosmos.

Would you prefer a recurrence of the primordial fireball, perhaps in the form of antimatter, with a repeat universe? Do you like the idea of an oscillating universe, which alternately falls inward and explodes outward, with similar (or perhaps much different) events each time in between? Do you believe that the matter in the universe will fall into a single point and disappear altogether from existence? The equations do not tell us which of these things is correct. Astronomers are doubtful that any breakthrough will ever be made in this respect. If we cannot find any more information to work with, then we have reached the limit of our knowledge about the universe.

Is this the final product of all our troubles: an unanswerable question? We might find comfort from the thought that we have been at this kind of impasse before. Astronomers have felt this sense of frustration and conquest many times. But always, just when it seems that science has reached a dead end, a whole new world of knowledge has been uncovered.

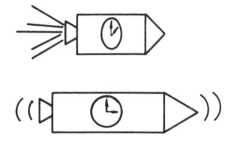

Suggested
Additional Reading

Born, Max. *Einstein's Theory of Relativity*. New York: Dover Publications, Inc., 1965.

Boslough, John. *Stephen Hawking's Universe*. New York: Avon Books, 1985.

Clark, Ronald W. *Einstein: The Life and Times*. New York: The World Publishing Company, 1971.

Einstein, Albert. *Relativity*. New York: Crown Publishers, 1961.

Hawking, Stephen. *A Brief History of Time*. New York: Bantam Books, 1988.

Herbert, Nick, Ph.D. *Faster than Light*. New York: NAL Penguin, Inc., 1988.

Jastrow, Robert. *God and the Astronomers*. New York: W. W. Norton & Co., Inc., 1978.

Sagan, Carl. *Cosmos*. New York: Random House, 1980.

Taylor, John G. *Black Holes*. New York: Avon Books, 1973.

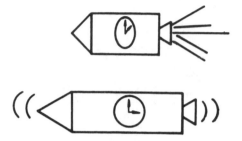

Index

A CATALOG OF SELECTED
DOVER BOOKS
IN ALL FIELDS OF INTEREST

A CATALOG OF SELECTED DOVER
BOOKS IN ALL FIELDS OF INTEREST

CONCERNING THE SPIRITUAL IN ART, Wassily Kandinsky. Pioneering work by father of abstract art. Thoughts on color theory, nature of art. Analysis of earlier masters. 12 illustrations. 80pp. of text. 5⅜ x 8½.   0-486-23411-8

CELTIC ART: The Methods of Construction, George Bain. Simple geometric techniques for making Celtic interlacements, spirals, Kells-type initials, animals, humans, etc. Over 500 illustrations. 160pp. 9 x 12. (Available in U.S. only.)   0-486-22923-8

AN ATLAS OF ANATOMY FOR ARTISTS, Fritz Schider. Most thorough reference work on art anatomy in the world. Hundreds of illustrations, including selections from works by Vesalius, Leonardo, Goya, Ingres, Michelangelo, others. 593 illustrations. 192pp. 7⅛ x 10¼.   0-486-20241-0

CELTIC HAND STROKE-BY-STROKE (Irish Half-Uncial from "The Book of Kells"): An Arthur Baker Calligraphy Manual, Arthur Baker. Complete guide to creating each letter of the alphabet in distinctive Celtic manner. Covers hand position, strokes, pens, inks, paper, more. Illustrated. 48pp. 8¼ x 11.   0-486-24336-2

EASY ORIGAMI, John Montroll. Charming collection of 32 projects (hat, cup, pelican, piano, swan, many more) specially designed for the novice origami hobbyist. Clearly illustrated easy-to-follow instructions insure that even beginning papercrafters will achieve successful results. 48pp. 8¼ x 11.   0-486-27298-2

BLOOMINGDALE'S ILLUSTRATED 1886 CATALOG: Fashions, Dry Goods and Housewares, Bloomingdale Brothers. Famed merchants' extremely rare catalog depicting about 1,700 products: clothing, housewares, firearms, dry goods, jewelry, more. Invaluable for dating, identifying vintage items. Also, copyright-free graphics for artists, designers. Co-published with Henry Ford Museum & Greenfield Village. 160pp. 8¼ x 11.   0-486-25780-0

THE ART OF WORLDLY WISDOM, Baltasar Gracian. "Think with the few and speak with the many," "Friends are a second existence," and "Be able to forget" are among this 1637 volume's 300 pithy maxims. A perfect source of mental and spiritual refreshment, it can be opened at random and appreciated either in brief or at length. 128pp. 5⅜ x 8½.   0-486-44034-6

JOHNSON'S DICTIONARY: A Modern Selection, Samuel Johnson (E. L. McAdam and George Milne, eds.). This modern version reduces the original 1755 edition's 2,300 pages of definitions and literary examples to a more manageable length, retaining the verbal pleasure and historical curiosity of the original. 480pp. 5¾₆ x 8¼.   0-486-44089-3

ADVENTURES OF HUCKLEBERRY FINN, Mark Twain, Illustrated by E. W. Kemble. A work of eternal richness and complexity, a source of ongoing critical debate, and a literary landmark, Twain's 1885 masterpiece about a barefoot boy's journey of self-discovery has enthralled readers around the world. This handsome clothbound reproduction of the first edition features all 174 of the original black-and-white illustrations. 368pp. 5⅜ x 8½.   0-486-44322-1

STICKLEY CRAFTSMAN FURNITURE CATALOGS, Gustav Stickley and L. & J. G. Stickley. Beautiful, functional furniture in two authentic catalogs from 1910. 594 illustrations, including 277 photos, show settles, rockers, armchairs, reclining chairs, bookcases, desks, tables. 183pp. 6½ x 9¼. 0-486-23838-5

AMERICAN LOCOMOTIVES IN HISTORIC PHOTOGRAPHS: 1858 to 1949, Ron Ziel (ed.). A rare collection of 126 meticulously detailed official photographs, called "builder portraits," of American locomotives that majestically chronicle the rise of steam locomotive power in America. Introduction. Detailed captions. xi+ 129pp. 9 x 12. 0-486-27393-8

AMERICA'S LIGHTHOUSES: An Illustrated History, Francis Ross Holland, Jr. Delightfully written, profusely illustrated fact-filled survey of over 200 American lighthouses since 1716. History, anecdotes, technological advances, more. 240pp. 8 x 10¾.
0-486-25576-X

TOWARDS A NEW ARCHITECTURE, Le Corbusier. Pioneering manifesto by founder of "International School." Technical and aesthetic theories, views of industry, economics, relation of form to function, "mass-production split" and much more. Profusely illustrated. 320pp. 6⅛ x 9¼. (Available in U.S. only.) 0-486-25023-7

HOW THE OTHER HALF LIVES, Jacob Riis. Famous journalistic record, exposing poverty and degradation of New York slums around 1900, by major social reformer. 100 striking and influential photographs. 233pp. 10 x 7⅞. 0-486-22012-5

FRUIT KEY AND TWIG KEY TO TREES AND SHRUBS, William M. Harlow. One of the handiest and most widely used identification aids. Fruit key covers 120 deciduous and evergreen species; twig key 160 deciduous species. Easily used. Over 300 photographs. 126pp. 5⅝ x 8½. 0-486-20511-8

COMMON BIRD SONGS, Dr. Donald J. Borror. Songs of 60 most common U.S. birds: robins, sparrows, cardinals, bluejays, finches, more–arranged in order of increasing complexity. Up to 9 variations of songs of each species.
Cassette and manual 0-486-99911-4

ORCHIDS AS HOUSE PLANTS, Rebecca Tyson Northen. Grow cattleyas and many other kinds of orchids–in a window, in a case, or under artificial light. 63 illustrations. 148pp. 5⅝ x 8½. 0-486-23261-1

MONSTER MAZES, Dave Phillips. Masterful mazes at four levels of difficulty. Avoid deadly perils and evil creatures to find magical treasures. Solutions for all 32 exciting illustrated puzzles. 48pp. 8¼ x 11. 0-486-26005-4

MOZART'S DON GIOVANNI (DOVER OPERA LIBRETTO SERIES), Wolfgang Amadeus Mozart. Introduced and translated by Ellen H. Bleiler. Standard Italian libretto, with complete English translation. Convenient and thoroughly portable–an ideal companion for reading along with a recording or the performance itself. Introduction. List of characters. Plot summary. 121pp. 5¼ x 8½. 0-486-24944-1

FRANK LLOYD WRIGHT'S DANA HOUSE, Donald Hoffmann. Pictorial essay of residential masterpiece with over 160 interior and exterior photos, plans, elevations, sketches and studies. 128pp. 9¼ x 10¾. 0-486-29120-0

THE CLARINET AND CLARINET PLAYING, David Pino. Lively, comprehensive work features suggestions about technique, musicianship, and musical interpretation, as well as guidelines for teaching, making your own reeds, and preparing for public performance. Includes an intriguing look at clarinet history. "A godsend," *The Clarinet,* Journal of the International Clarinet Society. Appendixes. 7 illus. 320pp. 5⅜ x 8½. 0-486-40270-3

HOLLYWOOD GLAMOR PORTRAITS, John Kobal (ed.). 145 photos from 1926-49. Harlow, Gable, Bogart, Bacall; 94 stars in all. Full background on photographers, technical aspects. 160pp. 8⅜ x 11¼. 0-486-23352-9

THE RAVEN AND OTHER FAVORITE POEMS, Edgar Allan Poe. Over 40 of the author's most memorable poems: "The Bells," "Ulalume," "Israfel," "To Helen," "The Conqueror Worm," "Eldorado," "Annabel Lee," many more. Alphabetic lists of titles and first lines. 64pp. 5³⁄₁₆ x 8¼. 0-486-26685-0

PERSONAL MEMOIRS OF U. S. GRANT, Ulysses Simpson Grant. Intelligent, deeply moving firsthand account of Civil War campaigns, considered by many the finest military memoirs ever written. Includes letters, historic photographs, maps and more. 528pp. 6⅛ x 9¼. 0-486-28587-1

ANCIENT EGYPTIAN MATERIALS AND INDUSTRIES, A. Lucas and J. Harris. Fascinating, comprehensive, thoroughly documented text describes this ancient civilization's vast resources and the processes that incorporated them in daily life, including the use of animal products, building materials, cosmetics, perfumes and incense, fibers, glazed ware, glass and its manufacture, materials used in the mummification process, and much more. 544pp. 6⅛ x 9¼. (Available in U.S. only.) 0-486-40446-3

RUSSIAN STORIES/RUSSKIE RASSKAZY: A Dual-Language Book, edited by Gleb Struve. Twelve tales by such masters as Chekhov, Tolstoy, Dostoevsky, Pushkin, others. Excellent word-for-word English translations on facing pages, plus teaching and study aids, Russian/English vocabulary, biographical/critical introductions, more. 416pp. 5⅜ x 8½. 0-486-26244-8

PHILADELPHIA THEN AND NOW: 60 Sites Photographed in the Past and Present, Kenneth Finkel and Susan Oyama. Rare photographs of City Hall, Logan Square, Independence Hall, Betsy Ross House, other landmarks juxtaposed with contemporary views. Captures changing face of historic city. Introduction. Captions. 128pp. 8¼ x 11. 0-486-25790-8

NORTH AMERICAN INDIAN LIFE: Customs and Traditions of 23 Tribes, Elsie Clews Parsons (ed.). 27 fictionalized essays by noted anthropologists examine religion, customs, government, additional facets of life among the Winnebago, Crow, Zuni, Eskimo, other tribes. 480pp. 6⅛ x 9¼. 0-486-27377-6

TECHNICAL MANUAL AND DICTIONARY OF CLASSICAL BALLET, Gail Grant. Defines, explains, comments on steps, movements, poses and concepts. 15-page pictorial section. Basic book for student, viewer. 127pp. 5⅜ x 8½. 0-486-21843-0

THE MALE AND FEMALE FIGURE IN MOTION: 60 Classic Photographic Sequences, Eadweard Muybridge. 60 true-action photographs of men and women walking, running, climbing, bending, turning, etc., reproduced from rare 19th-century masterpiece. vi + 121pp. 9 x 12. 0-486-24745-7

ANIMALS: 1,419 Copyright-Free Illustrations of Mammals, Birds, Fish, Insects, etc., Jim Harter (ed.). Clear wood engravings present, in extremely lifelike poses, over 1,000 species of animals. One of the most extensive pictorial sourcebooks of its kind. Captions. Index. 284pp. 9 x 12. 0-486-23766-4

1001 QUESTIONS ANSWERED ABOUT THE SEASHORE, N. J. Berrill and Jacquelyn Berrill. Queries answered about dolphins, sea snails, sponges, starfish, fishes, shore birds, many others. Covers appearance, breeding, growth, feeding, much more. 305pp. 5¼ x 8¼. 0-486-23366-9

ATTRACTING BIRDS TO YOUR YARD, William J. Weber. Easy-to-follow guide offers advice on how to attract the greatest diversity of birds: birdhouses, feeders, water and waterers, much more. 96pp. 5³⁄₁₆ x 8¼. 0-486-28927-3

MEDICINAL AND OTHER USES OF NORTH AMERICAN PLANTS: A Historical Survey with Special Reference to the Eastern Indian Tribes, Charlotte Erichsen-Brown. Chronological historical citations document 500 years of usage of plants, trees, shrubs native to eastern Canada, northeastern U.S. Also complete identifying information. 343 illustrations. 544pp. 6½ x 9¼. 0-486-25951-X

STORYBOOK MAZES, Dave Phillips. 23 stories and mazes on two-page spreads: Wizard of Oz, Treasure Island, Robin Hood, etc. Solutions. 64pp. 8¼ x 11. 0-486-23628-5

AMERICAN NEGRO SONGS: 230 Folk Songs and Spirituals, Religious and Secular, John W. Work. This authoritative study traces the African influences of songs sung and played by black Americans at work, in church, and as entertainment. The author discusses the lyric significance of such songs as "Swing Low, Sweet Chariot," "John Henry," and others and offers the words and music for 230 songs. Bibliography. Index of Song Titles. 272pp. 6½ x 9¼. 0-486-40271-1

MOVIE-STAR PORTRAITS OF THE FORTIES, John Kobal (ed.). 163 glamor, studio photos of 106 stars of the 1940s: Rita Hayworth, Ava Gardner, Marlon Brando, Clark Gable, many more. 176pp. 8⅜ x 11¼. 0-486-23546-7

YEKL and THE IMPORTED BRIDEGROOM AND OTHER STORIES OF YIDDISH NEW YORK, Abraham Cahan. Film Hester Street based on *Yekl* (1896). Novel, other stories among first about Jewish immigrants on N.Y.'s East Side. 240pp. 5⅜ x 8½. 0-486-22427-9

SELECTED POEMS, Walt Whitman. Generous sampling from *Leaves of Grass*. Twenty-four poems include "I Hear America Singing," "Song of the Open Road," "I Sing the Body Electric," "When Lilacs Last in the Dooryard Bloom'd," "O Captain! My Captain!"—all reprinted from an authoritative edition. Lists of titles and first lines. 128pp. 5³⁄₁₆ x 8¼. 0-486-26878-0

SONGS OF EXPERIENCE: Facsimile Reproduction with 26 Plates in Full Color, William Blake. 26 full-color plates from a rare 1826 edition. Includes "The Tyger," "London," "Holy Thursday," and other poems. Printed text of poems. 48pp. 5¼ x 7. 0-486-24636-1

THE BEST TALES OF HOFFMANN, E. T. A. Hoffmann. 10 of Hoffmann's most important stories: "Nutcracker and the King of Mice," "The Golden Flowerpot," etc. 458pp. 5⅜ x 8½. 0-486-21793-0

THE BOOK OF TEA, Kakuzo Okakura. Minor classic of the Orient: entertaining, charming explanation, interpretation of traditional Japanese culture in terms of tea ceremony. 94pp. 5⅜ x 8½. 0-486-20070-1

FRENCH STORIES/CONTES FRANÇAIS: A Dual-Language Book, Wallace Fowlie. Ten stories by French masters, Voltaire to Camus: "Micromegas" by Voltaire; "The Atheist's Mass" by Balzac; "Minuet" by de Maupassant; "The Guest" by Camus, six more. Excellent English translations on facing pages. Also French-English vocabulary list, exercises, more. 352pp. 5⅜ x 8½. 0-486-26443-2

CHICAGO AT THE TURN OF THE CENTURY IN PHOTOGRAPHS: 122 Historic Views from the Collections of the Chicago Historical Society, Larry A. Viskochil. Rare large-format prints offer detailed views of City Hall, State Street, the Loop, Hull House, Union Station, many other landmarks, circa 1904-1913. Introduction. Captions. Maps. 144pp. 9⅜ x 12¼. 0-486-24656-6

OLD BROOKLYN IN EARLY PHOTOGRAPHS, 1865-1929, William Lee Younger. Luna Park, Gravesend race track, construction of Grand Army Plaza, moving of Hotel Brighton, etc. 157 previously unpublished photographs. 165pp. 8⅞ x 11¾. 0-486-23587-4

THE MYTHS OF THE NORTH AMERICAN INDIANS, Lewis Spence. Rich anthology of the myths and legends of the Algonquins, Iroquois, Pawnees and Sioux, prefaced by an extensive historical and ethnological commentary. 36 illustrations. 480pp. 5⅜ x 8½. 0-486-25967-6

AN ENCYCLOPEDIA OF BATTLES: Accounts of Over 1,560 Battles from 1479 B.C. to the Present, David Eggenberger. Essential details of every major battle in recorded history from the first battle of Megiddo in 1479 B.C. to Grenada in 1984. List of Battle Maps. New Appendix covering the years 1967-1984. Index. 99 illustrations. 544pp. 6½ x 9¼. 0-486-24913-1

SAILING ALONE AROUND THE WORLD, Captain Joshua Slocum. First man to sail around the world, alone, in small boat. One of great feats of seamanship told in delightful manner. 67 illustrations. 294pp. 5⅜ x 8½. 0-486-20326-3

ANARCHISM AND OTHER ESSAYS, Emma Goldman. Powerful, penetrating, prophetic essays on direct action, role of minorities, prison reform, puritan hypocrisy, violence, etc. 271pp. 5⅜ x 8½. 0-486-22484-8

MYTHS OF THE HINDUS AND BUDDHISTS, Ananda K. Coomaraswamy and Sister Nivedita. Great stories of the epics; deeds of Krishna, Shiva, taken from puranas, Vedas, folk tales; etc. 32 illustrations. 400pp. 5⅜ x 8½. 0-486-21759-0

MY BONDAGE AND MY FREEDOM, Frederick Douglass. Born a slave, Douglass became outspoken force in antislavery movement. The best of Douglass' autobiographies. Graphic description of slave life. 464pp. 5⅜ x 8½. 0-486-22457-0

FOLLOWING THE EQUATOR: A Journey Around the World, Mark Twain. Fascinating humorous account of 1897 voyage to Hawaii, Australia, India, New Zealand, etc. Ironic, bemused reports on peoples, customs, climate, flora and fauna, politics, much more. 197 illustrations. 720pp. 5⅜ x 8½. 0-486-26113-1

THE PEOPLE CALLED SHAKERS, Edward D. Andrews. Definitive study of Shakers: origins, beliefs, practices, dances, social organization, furniture and crafts, etc. 33 illustrations. 351pp. 5⅜ x 8½. 0-486-21081-2

THE MYTHS OF GREECE AND ROME, H. A. Guerber. A classic of mythology, generously illustrated, long prized for its simple, graphic, accurate retelling of the principal myths of Greece and Rome, and for its commentary on their origins and significance. With 64 illustrations by Michelangelo, Raphael, Titian, Rubens, Canova, Bernini and others. 480pp. 5⅜ x 8½. 0-486-27584-1

PSYCHOLOGY OF MUSIC, Carl E. Seashore. Classic work discusses music as a medium from psychological viewpoint. Clear treatment of physical acoustics, auditory apparatus, sound perception, development of musical skills, nature of musical feeling, host of other topics. 88 figures. 408pp. $5\frac{3}{8}$ x $8\frac{1}{2}$. 0-486-21851-1

LIFE IN ANCIENT EGYPT, Adolf Erman. Fullest, most thorough, detailed older account with much not in more recent books, domestic life, religion, magic, medicine, commerce, much more. Many illustrations reproduce tomb paintings, carvings, hieroglyphs, etc. 597pp. $5\frac{3}{8}$ x $8\frac{1}{2}$. 0-486-22632-8

SUNDIALS, Their Theory and Construction, Albert Waugh. Far and away the best, most thorough coverage of ideas, mathematics concerned, types, construction, adjusting anywhere. Simple, nontechnical treatment allows even children to build several of these dials. Over 100 illustrations. 230pp. $5\frac{3}{8}$ x $8\frac{1}{2}$. 0-486-22947-5

THEORETICAL HYDRODYNAMICS, L. M. Milne-Thomson. Classic exposition of the mathematical theory of fluid motion, applicable to both hydrodynamics and aerodynamics. Over 600 exercises. 768pp. $6\frac{1}{8}$ x $9\frac{1}{4}$. 0-486-68970-0

OLD-TIME VIGNETTES IN FULL COLOR, Carol Belanger Grafton (ed.). Over 390 charming, often sentimental illustrations, selected from archives of Victorian graphics–pretty women posing, children playing, food, flowers, kittens and puppies, smiling cherubs, birds and butterflies, much more. All copyright-free. 48pp. $9\frac{1}{4}$ x $12\frac{1}{4}$. 0-486-27269-9

PERSPECTIVE FOR ARTISTS, Rex Vicat Cole. Depth, perspective of sky and sea, shadows, much more, not usually covered. 391 diagrams, 81 reproductions of drawings and paintings. 279pp. $5\frac{3}{8}$ x $8\frac{1}{2}$. 0-486-22487-2

DRAWING THE LIVING FIGURE, Joseph Sheppard. Innovative approach to artistic anatomy focuses on specifics of surface anatomy, rather than muscles and bones. Over 170 drawings of live models in front, back and side views, and in widely varying poses. Accompanying diagrams. 177 illustrations. Introduction. Index. 144pp. $8\frac{3}{8}$ x$11\frac{1}{4}$. 0-486-26723-7

GOTHIC AND OLD ENGLISH ALPHABETS: 100 Complete Fonts, Dan X. Solo. Add power, elegance to posters, signs, other graphics with 100 stunning copyright-free alphabets: Blackstone, Dolbey, Germania, 97 more–including many lower-case, numerals, punctuation marks. 104pp. $8\frac{1}{8}$ x 11. 0-486-24695-7

THE BOOK OF WOOD CARVING, Charles Marshall Sayers. Finest book for beginners discusses fundamentals and offers 34 designs. "Absolutely first rate . . . well thought out and well executed."–E. J. Tangerman. 118pp. $7\frac{3}{4}$ x $10\frac{5}{8}$. 0-486-23654-4

ILLUSTRATED CATALOG OF CIVIL WAR MILITARY GOODS: Union Army Weapons, Insignia, Uniform Accessories, and Other Equipment, Schuyler, Hartley, and Graham. Rare, profusely illustrated 1846 catalog includes Union Army uniform and dress regulations, arms and ammunition, coats, insignia, flags, swords, rifles, etc. 226 illustrations. 160pp. 9 x 12. 0-486-24939-5

WOMEN'S FASHIONS OF THE EARLY 1900s: An Unabridged Republication of "New York Fashions, 1909," National Cloak & Suit Co. Rare catalog of mail-order fashions documents women's and children's clothing styles shortly after the turn of the century. Captions offer full descriptions, prices. Invaluable resource for fashion, costume historians. Approximately 725 illustrations. 128pp. $8\frac{3}{8}$ x $11\frac{1}{4}$. 0-486-27276-1

CATALOG OF DOVER BOOKS

HOW TO DO BEADWORK, Mary White. Fundamental book on craft from simple projects to five-bead chains and woven works. 106 illustrations. 142pp. 5⅜ x 8. 0-486-20697-1

THE 1912 AND 1915 GUSTAV STICKLEY FURNITURE CATALOGS, Gustav Stickley. With over 200 detailed illustrations and descriptions, these two catalogs are essential reading and reference materials and identification guides for Stickley furniture. Captions cite materials, dimensions and prices. 112pp. 6½ x 9¼. 0-486-26676-1

EARLY AMERICAN LOCOMOTIVES, John H. White, Jr. Finest locomotive engravings from early 19th century: historical (1804–74), main-line (after 1870), special, foreign, etc. 147 plates. 142pp. 11⅝ x 8¼. 0-486-22772-3

LITTLE BOOK OF EARLY AMERICAN CRAFTS AND TRADES, Peter Stockham (ed.). 1807 children's book explains crafts and trades: baker, hatter, cooper, potter, and many others. 23 copperplate illustrations. 140pp. 4⅝ x 6. 0-486-23336-7

VICTORIAN FASHIONS AND COSTUMES FROM HARPER'S BAZAR, 1867–1898, Stella Blum (ed.). Day costumes, evening wear, sports clothes, shoes, hats, other accessories in over 1,000 detailed engravings. 320pp. 9⅜ x 12¼. 0-486-22990-4

THE LONG ISLAND RAIL ROAD IN EARLY PHOTOGRAPHS, Ron Ziel. Over 220 rare photos, informative text document origin (1844) and development of rail service on Long Island. Vintage views of early trains, locomotives, stations, passengers, crews, much more. Captions. 8⅞ x 11¾. 0-486-26301-0

VOYAGE OF THE LIBERDADE, Joshua Slocum. Great 19th-century mariner's thrilling, first-hand account of the wreck of his ship off South America, the 35-foot boat he built from the wreckage, and its remarkable voyage home. 128pp. 5⅜ x 8½. 0-486-40022-0

TEN BOOKS ON ARCHITECTURE, Vitruvius. The most important book ever written on architecture. Early Roman aesthetics, technology, classical orders, site selection, all other aspects. Morgan translation. 331pp. 5⅜ x 8½. 0-486-20645-9

THE HUMAN FIGURE IN MOTION, Eadweard Muybridge. More than 4,500 stopped-action photos, in action series, showing undraped men, women, children jumping, lying down, throwing, sitting, wrestling, carrying, etc. 390pp. 7⅞ x 10⅝. 0-486-20204-6 Clothbd.

TREES OF THE EASTERN AND CENTRAL UNITED STATES AND CANADA, William M. Harlow. Best one-volume guide to 140 trees. Full descriptions, woodlore, range, etc. Over 600 illustrations. Handy size. 288pp. 4½ x 6⅜. 0-486-20395-6

GROWING AND USING HERBS AND SPICES, Milo Miloradovich. Versatile handbook provides all the information needed for cultivation and use of all the herbs and spices available in North America. 4 illustrations. Index. Glossary. 236pp. 5⅜ x 8½. 0-486-25058-X

BIG BOOK OF MAZES AND LABYRINTHS, Walter Shepherd. 50 mazes and labyrinths in all–classical, solid, ripple, and more–in one great volume. Perfect inexpensive puzzler for clever youngsters. Full solutions. 112pp. 8⅛ x 11. 0-486-22951-3

PIANO TUNING, J. Cree Fischer. Clearest, best book for beginner, amateur. Simple repairs, raising dropped notes, tuning by easy method of flattened fifths. No previous skills needed. 4 illustrations. 201pp. 5⅜ x 8½. 0-486-23267-0

HINTS TO SINGERS, Lillian Nordica. Selecting the right teacher, developing confidence, overcoming stage fright, and many other important skills receive thoughtful discussion in this indispensible guide, written by a world-famous diva of four decades' experience. 96pp. 5⅜ x 8½. 0-486-40094-8

THE COMPLETE NONSENSE OF EDWARD LEAR, Edward Lear. All nonsense limericks, zany alphabets, Owl and Pussycat, songs, nonsense botany, etc., illustrated by Lear. Total of 320pp. 5⅜ x 8½. (Available in U.S. only.) 0-486-20167-8

VICTORIAN PARLOUR POETRY: An Annotated Anthology, Michael R. Turner. 117 gems by Longfellow, Tennyson, Browning, many lesser-known poets. "The Village Blacksmith," "Curfew Must Not Ring Tonight," "Only a Baby Small," dozens more, often difficult to find elsewhere. Index of poets, titles, first lines. xxiii + 325pp. 5⅜ x 8¼. 0-486-27044-0

DUBLINERS, James Joyce. Fifteen stories offer vivid, tightly focused observations of the lives of Dublin's poorer classes. At least one, "The Dead," is considered a masterpiece. Reprinted complete and unabridged from standard edition. 160pp. 5³⁄₁₆ x 8¼. 0-486-26870-5

GREAT WEIRD TALES: 14 Stories by Lovecraft, Blackwood, Machen and Others, S. T. Joshi (ed.). 14 spellbinding tales, including "The Sin Eater," by Fiona McLeod, "The Eye Above the Mantel," by Frank Belknap Long, as well as renowned works by R. H. Barlow, Lord Dunsany, Arthur Machen, W. C. Morrow and eight other masters of the genre. 256pp. 5⅜ x 8½. (Available in U.S. only.) 0-486-40436-6

THE BOOK OF THE SACRED MAGIC OF ABRAMELIN THE MAGE, translated by S. MacGregor Mathers. Medieval manuscript of ceremonial magic. Basic document in Aleister Crowley, Golden Dawn groups. 268pp. 5⅜ x 8½.

0-486-23211-5

THE BATTLES THAT CHANGED HISTORY, Fletcher Pratt. Eminent historian profiles 16 crucial conflicts, ancient to modern, that changed the course of civilization. 352pp. 5⅜ x 8½. 0-486-41129-X

NEW RUSSIAN-ENGLISH AND ENGLISH-RUSSIAN DICTIONARY, M. A. O'Brien. This is a remarkably handy Russian dictionary, containing a surprising amount of information, including over 70,000 entries. 366pp. 4½ x 6⅛. 0-486-20208-9

NEW YORK IN THE FORTIES, Andreas Feininger. 162 brilliant photographs by the well-known photographer, formerly with *Life* magazine. Commuters, shoppers, Times Square at night, much else from city at its peak. Captions by John von Hartz. 181pp. 9¼ x 10¾. 0-486-23585-8

INDIAN SIGN LANGUAGE, William Tomkins. Over 525 signs developed by Sioux and other tribes. Written instructions and diagrams. Also 290 pictographs. 111pp. 6⅛ x 9¼. 0-486-22029-X

ANATOMY: A Complete Guide for Artists, Joseph Sheppard. A master of figure drawing shows artists how to render human anatomy convincingly. Over 460 illustrations. 224pp. 8⅜ x 11¼. 0-486-27279-6

MEDIEVAL CALLIGRAPHY: Its History and Technique, Marc Drogin. Spirited history, comprehensive instruction manual covers 13 styles (ca. 4th century through 15th). Excellent photographs; directions for duplicating medieval techniques with modern tools. 224pp. 8⅜ x 11¼. 0-486-26142-5

DRIED FLOWERS: How to Prepare Them, Sarah Whitlock and Martha Rankin. Complete instructions on how to use silica gel, meal and borax, perlite aggregate, sand and borax, glycerine and water to create attractive permanent flower arrangements. 12 illustrations. 32pp. 5⅜ x 8½. 0-486-21802-3

EASY-TO-MAKE BIRD FEEDERS FOR WOODWORKERS, Scott D. Campbell. Detailed, simple-to-use guide for designing, constructing, caring for and using feeders. Text, illustrations for 12 classic and contemporary designs. 96pp. 5⅜ x 8½. 0-486-25847-5

THE COMPLETE BOOK OF BIRDHOUSE CONSTRUCTION FOR WOODWORKERS, Scott D. Campbell. Detailed instructions, illustrations, tables. Also data on bird habitat and instinct patterns. Bibliography. 3 tables. 63 illustrations in 15 figures. 48pp. 5¼ x 8¼. 0-486-24407-5

SCOTTISH WONDER TALES FROM MYTH AND LEGEND, Donald A. Mackenzie. 16 lively tales tell of giants rumbling down mountainsides, of a magic wand that turns stone pillars into warriors, of gods and goddesses, evil hags, powerful forces and more. 240pp. 5⅜ x 8½. 0-486-29677-6

THE HISTORY OF UNDERCLOTHES, C. Willett Cunnington and Phyllis Cunnington. Fascinating, well-documented survey covering six centuries of English undergarments, enhanced with over 100 illustrations: 12th-century laced-up bodice, footed long drawers (1795), 19th-century bustles, l9th-century corsets for men, Victorian "bust improvers," much more. 272pp. 5⅜ x 8¼. 0-486-27124-2

ARTS AND CRAFTS FURNITURE: The Complete Brooks Catalog of 1912, Brooks Manufacturing Co. Photos and detailed descriptions of more than 150 now very collectible furniture designs from the Arts and Crafts movement depict davenports, settees, buffets, desks, tables, chairs, bedsteads, dressers and more, all built of solid, quarter-sawed oak. Invaluable for students and enthusiasts of antiques, Americana and the decorative arts. 80pp. 6½ x 9¼. 0-486-27471-3

WILBUR AND ORVILLE: A Biography of the Wright Brothers, Fred Howard. Definitive, crisply written study tells the full story of the brothers' lives and work. A vividly written biography, unparalleled in scope and color, that also captures the spirit of an extraordinary era. 560pp. 6⅛ x 9¼. 0-486-40297-5

THE ARTS OF THE SAILOR: Knotting, Splicing and Ropework, Hervey Garrett Smith. Indispensable shipboard reference covers tools, basic knots and useful hitches; handsewing and canvas work, more. Over 100 illustrations. Delightful reading for sea lovers. 256pp. 5⅜ x 8½. 0-486-26440-8

FRANK LLOYD WRIGHT'S FALLINGWATER: The House and Its History, Second, Revised Edition, Donald Hoffmann. A total revision–both in text and illustrations–of the standard document on Fallingwater, the boldest, most personal architectural statement of Wright's mature years, updated with valuable new material from the recently opened Frank Lloyd Wright Archives. "Fascinating"–*The New York Times*. 116 illustrations. 128pp. 9¼ x 10¾. 0-486-27430-6

PHOTOGRAPHIC SKETCHBOOK OF THE CIVIL WAR, Alexander Gardner. 100 photos taken on field during the Civil War. Famous shots of Manassas Harper's Ferry, Lincoln, Richmond, slave pens, etc. 244pp. 10⅝ x 8¼. 0-486-22731-6

FIVE ACRES AND INDEPENDENCE, Maurice G. Kains. Great back-to-the-land classic explains basics of self-sufficient farming. The one book to get. 95 illustrations. 397pp. 5⅜ x 8½. 0-486-20974-1

A MODERN HERBAL, Margaret Grieve. Much the fullest, most exact, most useful compilation of herbal material. Gigantic alphabetical encyclopedia, from aconite to zedoary, gives botanical information, medical properties, folklore, economic uses, much else. Indispensable to serious reader. 161 illustrations. 888pp. 6½ x 9¼. 2-vol. set. (Available in U.S. only.)   Vol. I: 0-486-22798-7   Vol. II: 0-486-22799-5

HIDDEN TREASURE MAZE BOOK, Dave Phillips. Solve 34 challenging mazes accompanied by heroic tales of adventure. Evil dragons, people-eating plants, bloodthirsty giants, many more dangerous adversaries lurk at every twist and turn. 34 mazes, stories, solutions. 48pp. 8¼ x 11.   0-486-24566-7

LETTERS OF W. A. MOZART, Wolfgang A. Mozart. Remarkable letters show bawdy wit, humor, imagination, musical insights, contemporary musical world; includes some letters from Leopold Mozart. 276pp. 5⅜ x 8½.   0-486-22859-2

BASIC PRINCIPLES OF CLASSICAL BALLET, Agrippina Vaganova. Great Russian theoretician, teacher explains methods for teaching classical ballet. 118 illustrations. 175pp. 5⅜ x 8½.   0-486-22036-2

THE JUMPING FROG, Mark Twain. Revenge edition. The original story of The Celebrated Jumping Frog of Calaveras County, a hapless French translation, and Twain's hilarious "retranslation" from the French. 12 illustrations. 66pp. 5⅜ x 8½.
                                                              0-486-22686-7

BEST REMEMBERED POEMS, Martin Gardner (ed.). The 126 poems in this superb collection of 19th- and 20th-century British and American verse range from Shelley's "To a Skylark" to the impassioned "Renascence" of Edna St. Vincent Millay and to Edward Lear's whimsical "The Owl and the Pussycat." 224pp. 5⅜ x 8½.
                                                              0-486-27165-X

COMPLETE SONNETS, William Shakespeare. Over 150 exquisite poems deal with love, friendship, the tyranny of time, beauty's evanescence, death and other themes in language of remarkable power, precision and beauty. Glossary of archaic terms. 80pp. 5³⁄₁₆ x 8¼.   0-486-26686-9

HISTORIC HOMES OF THE AMERICAN PRESIDENTS, Second, Revised Edition, Irvin Haas. A traveler's guide to American Presidential homes, most open to the public, depicting and describing homes occupied by every American President from George Washington to George Bush. With visiting hours, admission charges, travel routes. 175 photographs. Index. 160pp. 8¼ x 11.   0-486-26751-2

THE WIT AND HUMOR OF OSCAR WILDE, Alvin Redman (ed.). More than 1,000 ripostes, paradoxes, wisecracks: Work is the curse of the drinking classes; I can resist everything except temptation; etc. 258pp. 5⅜ x 8½.   0-486-20602-5

SHAKESPEARE LEXICON AND QUOTATION DICTIONARY, Alexander Schmidt. Full definitions, locations, shades of meaning in every word in plays and poems. More than 50,000 exact quotations. 1,485pp. 6½ x 9¼. 2-vol. set.
                    Vol. 1: 0-486-22726-X   Vol. 2: 0-486-22727-8

SELECTED POEMS, Emily Dickinson. Over 100 best-known, best-loved poems by one of America's foremost poets, reprinted from authoritative early editions. No comparable edition at this price. Index of first lines. 64pp. 5³⁄₁₆ x 8¼. 0-486-26466-1

THE INSIDIOUS DR. FU-MANCHU, Sax Rohmer. The first of the popular mystery series introduces a pair of English detectives to their archnemesis, the diabolical Dr. Fu-Manchu. Flavorful atmosphere, fast-paced action, and colorful characters enliven this classic of the genre. 208pp. 5³⁄₁₆ x 8¼.   0-486-29898-1

THE MALLEUS MALEFICARUM OF KRAMER AND SPRENGER, translated by Montague Summers. Full text of most important witchhunter's "bible," used by both Catholics and Protestants. 278pp. 6⅝ x 10.   0-486-22802-9

SPANISH STORIES/CUENTOS ESPAÑOLES: A Dual-Language Book, Angel Flores (ed.). Unique format offers 13 great stories in Spanish by Cervantes, Borges, others. Faithful English translations on facing pages. 352pp. 5⅜ x 8½.

0-486-25399-6

GARDEN CITY, LONG ISLAND, IN EARLY PHOTOGRAPHS, 1869–1919, Mildred H. Smith. Handsome treasury of 118 vintage pictures, accompanied by carefully researched captions, document the Garden City Hotel fire (1899), the Vanderbilt Cup Race (1908), the first airmail flight departing from the Nassau Boulevard Aerodrome (1911), and much more. 96pp. 8⅞ x 11¾.   0-486-40669-5

OLD QUEENS, N.Y., IN EARLY PHOTOGRAPHS, Vincent F. Seyfried and William Asadorian. Over 160 rare photographs of Maspeth, Jamaica, Jackson Heights, and other areas. Vintage views of DeWitt Clinton mansion, 1939 World's Fair and more. Captions. 192pp. 8⅞ x 11.   0-486-26358-4

CAPTURED BY THE INDIANS: 15 Firsthand Accounts, 1750-1870, Frederick Drimmer. Astounding true historical accounts of grisly torture, bloody conflicts, relentless pursuits, miraculous escapes and more, by people who lived to tell the tale. 384pp. 5⅜ x 8½.   0-486-24901-8

THE WORLD'S GREAT SPEECHES (Fourth Enlarged Edition), Lewis Copeland, Lawrence W. Lamm, and Stephen J. McKenna. Nearly 300 speeches provide public speakers with a wealth of updated quotes and inspiration–from Pericles' funeral oration and William Jennings Bryan's "Cross of Gold Speech" to Malcolm X's powerful words on the Black Revolution and Earl of Spenser's tribute to his sister, Diana, Princess of Wales. 944pp. 5⅜ x 8⅜.   0-486-40903-1

THE BOOK OF THE SWORD, Sir Richard F. Burton. Great Victorian scholar/adventurer's eloquent, erudite history of the "queen of weapons"–from prehistory to early Roman Empire. Evolution and development of early swords, variations (sabre, broadsword, cutlass, scimitar, etc.), much more. 336pp. 6⅛ x 9¼.

0-486-25434-8

AUTOBIOGRAPHY: The Story of My Experiments with Truth, Mohandas K. Gandhi. Boyhood, legal studies, purification, the growth of the Satyagraha (nonviolent protest) movement. Critical, inspiring work of the man responsible for the freedom of India. 480pp. 5⅜ x 8½. (Available in U.S. only.)   0-486-24593-4

CELTIC MYTHS AND LEGENDS, T. W. Rolleston. Masterful retelling of Irish and Welsh stories and tales. Cuchulain, King Arthur, Deirdre, the Grail, many more. First paperback edition. 58 full-page illustrations. 512pp. 5⅜ x 8½.   0-486-26507-2

THE PRINCIPLES OF PSYCHOLOGY, William James. Famous long course complete, unabridged. Stream of thought, time perception, memory, experimental methods; great work decades ahead of its time. 94 figures. 1,391pp. 5⅜ x 8½. 2-vol. set.
Vol. I: 0-486-20381-6   Vol. II: 0-486-20382-4

THE WORLD AS WILL AND REPRESENTATION, Arthur Schopenhauer. Definitive English translation of Schopenhauer's life work, correcting more than 1,000 errors, omissions in earlier translations. Translated by E. F. J. Payne. Total of 1,269pp. 5⅜ x 8½. 2-vol. set.   Vol. 1: 0-486-21761-2   Vol. 2: 0-486-21762-0

CATALOG OF DOVER BOOKS

LIGHT AND SHADE: A Classic Approach to Three-Dimensional Drawing, Mrs. Mary P. Merrifield. Handy reference clearly demonstrates principles of light and shade by revealing effects of common daylight, sunshine, and candle or artificial light on geometrical solids. 13 plates. 64pp. 5⅜ x 8½.　　　　　　0-486-44143-1

ASTROLOGY AND ASTRONOMY: A Pictorial Archive of Signs and Symbols, Ernst and Johanna Lehner. Treasure trove of stories, lore, and myth, accompanied by more than 300 rare illustrations of planets, the Milky Way, signs of the zodiac, comets, meteors, and other astronomical phenomena. 192pp. 8⅜ x 11.

0-486-43981-X

JEWELRY MAKING: Techniques for Metal, Tim McCreight. Easy-to-follow instructions and carefully executed illustrations describe tools and techniques, use of gems and enamels, wire inlay, casting, and other topics. 72 line illustrations and diagrams. 176pp. 8¼ x 10⅞.　　　　　　0-486-44043-5

MAKING BIRDHOUSES: Easy and Advanced Projects, Gladstone Califf. Easy-to-follow instructions include diagrams for everything from a one-room house for bluebirds to a forty-two-room structure for purple martins. 56 plates; 4 figures. 80pp. 8¾ x 6¾.　　　　　　0-486-44183-0

LITTLE BOOK OF LOG CABINS: How to Build and Furnish Them, William S. Wicks. Handy how-to manual, with instructions and illustrations for building cabins in the Adirondack style, fireplaces, stairways, furniture, beamed ceilings, and more. 102 line drawings. 96pp. 8¾ x 6¾.　　　　　　0-486-44259-4

THE SEASONS OF AMERICA PAST, Eric Sloane. From "sugaring time" and strawberry picking to Indian summer and fall harvest, a whole year's activities described in charming prose and enhanced with 79 of the author's own illustrations. 160pp. 8¼ x 11.　　　　　　0-486-44220-9

THE METROPOLIS OF TOMORROW, Hugh Ferriss. Generous, prophetic vision of the metropolis of the future, as perceived in 1929. Powerful illustrations of towering structures, wide avenues, and rooftop parks—all features in many of today's modern cities. 59 illustrations. 144pp. 8¼ x 11.　　　　　　0-486-43727-2

THE PATH TO ROME, Hilaire Belloc. This 1902 memoir abounds in lively vignettes from a vanished time, recounting a pilgrimage on foot across the Alps and Apennines in order to "see all Europe which the Christian Faith has saved." 77 of the author's original line drawings complement his sparkling prose. 272pp. 5⅜ x 8½.

0-486-44001-X

THE HISTORY OF RASSELAS: Prince of Abissinia, Samuel Johnson. Distinguished English writer attacks eighteenth-century optimism and man's unrealistic estimates of what life has to offer. 112pp. 5⅜ x 8½.　　　　　　0-486-44094-X

A VOYAGE TO ARCTURUS, David Lindsay. A brilliant flight of pure fancy, where wild creatures crowd the fantastic landscape and demented torturers dominate victims with their bizarre mental powers. 272pp. 5⅜ x 8½.　　　　　　0-486-44198-9

Paperbound unless otherwise indicated. Available at your book dealer, online at **www.doverpublications.com**, or by writing to Dept. GI, Dover Publications, Inc., 31 East 2nd Street, Mineola, NY 11501. For current price information or for free catalogs (please indicate field of interest), write to Dover Publications or log on to **www.doverpublications.com** and see every Dover book in print. Dover publishes more than 500 books each year on science, elementary and advanced mathematics, biology, music, art, literary history, social sciences, and other areas.